AF478582

MODERN CELL BIOLOGY

Volume 1

MODERN CELL BIOLOGY

Volume 1

Series Editor

Birgit H. Satir

Department of Anatomy
Albert Einstein College of Medicine
Bronx, NY 10461

Alan R. Liss, Inc., New York

ISBN 0-8451-3300-4 • ISSN 0745-3000

Contents

Contributors

Richard G.W. Anderson [1]
Department of Cell Biology, University of Texas Health Science Center, Dallas, TX 75235

Leo T. Furcht [53]
Department of Laboratory Medicine and Pathology, University of Minnesota, Minneapolis, MN 55455

Jerry Kaplan [1]
Department of Pathology, University of Utah School of Medicine, Salt Lake City, UT 84132

Alfred Maelicke [171]
Max-Planck-Institut für Ernährungsphysiologie, Rheinlanddamm 201, D-4600 Dortmund, West Germany

Heino Prinz [171]
Max-Planck-Institut für Ernährungsphysiologie, Rheinlanddamm 201, D-4600 Dortmund, West Germany

Enrique Rodriguez-Boulan [119]
Department of Pathology, State University of New York, Downstate Medical Center, Brooklyn, NY 11203

The boldface number in brackets following each author's name is the opening page number of that author's article.

Foreword

The goal of this series is to provide researchers in cell and molecular biology with comprehensive articles that review areas of current significance and raise pertinent questions about future directions. We seek to publish articles that look to the future but provide lasting value. This is a need that is not being fully satisfied by any of the existing publications.

These are exciting times for the cell biologist. Our own advances and insights are being stimulated and enriched not only by research done by fellow cell biologists, but by research from our colleagues in such diverse but increasingly interconnected fields as molecular biology, developmental biology, biophysics, immunology, genetics, bioengineering, etc. This gives a particular urgency to the task of creating a review series truly reflective of the continuing evolution of our understanding of the biology of the cell. To help us do this we have gathered together an advisory board that is broadly based in both techniques and interests, as well as international in its composition, to reflect a diversity of breadth and depth that this series seeks to provide.

The papers in this series will be critical reviews of specific areas in cell biology and related fields, including personal perspectives, and general reviews related to the boundaries of the field or the application of new methods to cell biological problems. From time to time, a single volume will be organized as an in depth exploration of an area of importance in cell biology. An example of this is Volume 2, Spatial Organization of Eukaryotic Cells, a special volume in honor of one of the distinguished leaders in the field, Keith R. Porter.

We hope readers find these reviews both useful and exciting.

Birgit Satir
December 1982

Modern Cell Biology, 1:1–52

Receptor-Mediated Endocytosis

Richard G. W. Anderson and Jerry Kaplan

From the Department of Cell Biology, University of Texas Health Science Center, Dallas, Texas 75235 (R.G.W.A.), and the Department of Pathology, University of Utah School of Medicine, Salt Lake City, Utah 84132 (J.K.)

I. INTRODUCTION

It has been nearly 90 years since Overton [1] first carried out his classical studies on the permeability properties of cellular membranes. Many investigators have since confirmed that the lipid composition of biological membranes renders them impermeable to either highly charged molecules or to molecules larger than 400 daltons [2]. Only if the membrane is altered through the insertion of specific types of membrane molecules, which are usually proteins, is the cell able to selectively take up extracellular molecules that ordinarily are unable to penetrate the lipid bilayer. It is this event that transforms the membrane into a selective barrier. In addition, cells have developed several different mechanisms for internalizing portions of the extracellular environment and to deliver the trapped contents to various compartments within the cell. This process, called "endocytosis," is adapted either to ingest small portions of the extracellular environment (pinocytosis) or to ingest large, particulate material (phagocytosis). Although the two processes achieve the same end, the internalization of material from the extracellular space, the physiologic mechanisms involved are quite distinct [3].

In this review, we wish to restrict our discussion to the endocytosis of molecular material (pinocytosis), a process that requires the invagination of small regions of the surface membrane to form a vesicular compartment that contains a portion of extracellular fluid. With few exceptions (see Endocytosis), this type of endocytosis is a constitutive process that takes place at a constant rate for a given cell under defined environmental conditions. Steinman et al [4–6] used horseradish peroxidase as a molecular marker to measure the rate of endocytosis in macrophages and L cells in culture. They found that for both cell types, 10^6 cells internalize 10^{-4} ml of culture media each hour. In the same time interval, macrophages internalize 186% and L cells 54% of their cell surface [6]. They also found that among different fibroblasts, the rate of endocytosis is approximately the same [5], although the growth conditions of the cells can influence this rate.

The endocytic process displayed by these cells lacks specificity and is inefficient. Molecules are internalized in proportion to their concentration in the extracellular environment without regard to their metabolic importance to the cell. To overcome this constraint, cells have devised ways to increase the effectiveness of endocytosis. They synthesize receptors that function in the plasma membrane to recognize, concentrate, and internalize by endocytosis specific groups of extracellular macromolecules. The coupling of membrane receptor activity to macromolecule internalization is called receptor-mediated endocytosis (RME). Over 20 different macromolecules or macromolecular complexes have now been shown to be internalized by RME [7, 8]. The diversity of this class of molecules, from transport proteins to hormones, suggests that the cell utilizes RME to meet a variety of different physiological needs.

Receptor-mediated endocytosis is a multiphase process. Specific cell surface receptors must be located on the cell surface. In addition, the cell must be able to carry out endocytosis, a process that often involves differentiated regions of the cell surface. The receptors must be located over those regions of membrane that are undergoing endocytosis to achieve efficient uptake of the bound molecule. Once inside the cell, the contents of the endocytic vesicle and the receptors must be segregated so that receptors and their specific ligand can be delivered to separate intracellular targets. Often two or more molecules can enter the cell through the same endocytic vesicle. They too must be segregated so that they can eventually go to different intracellular compartments.

This highly organized cellular process has been studied by many investigators using several different cell systems. Necessarily, this review can neither cover every aspect of endocytosis nor comprehensively review receptor biology [see 3, 7, 9–12 for reviews]. Even an exhaustive compilation of the literature on RME will not be achieved. Our purpose in this review is to bring together diverse experimental findings and use this information to construct various models that will be useful for identifying the salient problems in this field, and we hope, to indicate new directions for future research.

II. RECEPTORS

A. Definition

Cell surface receptors are able to recognize and bind with high affinity specific subsets of extracellular macromolecules; furthermore, the binding step usually elicits a cellular response. In the case of those receptors involved in RME, a major response is the internalization of the ligand. This may be preceded by the generation of a signal that alters cellular metabolism (eg, polypeptide hormone receptors), or the internalized ligand may be utilized by the cell for specific metabolic needs. In either case, ligand binding is a physiologically important event.

Ligand-receptor interaction is specific and involves only one family of homologous extracellular molecules and one set of plasma membrane proteins. These receptors usually have been found to be a single protein or protein-protein complex [eg, see 13–17]. Moreover, the binding of the specific ligand depends on characteristic ionic and pH conditions. Ligand-receptor interactions have often been further defined by assessing how the specific modification of either the receptor or the ligand inactivates the binding step.

These receptors can therefore be defined by their molecular properties, the conditions for ligand binding, and their ability to mediate a specific physiologic event. It is this last property that has usually been responsible for their initial detection. For example, LDL receptors were discovered because of their ability to regulate intracellular cholesterol metabolism [28, 29]. Similarily, the asialoglycoprotein receptor [30] and the lysosomal enzyme receptor [31], to mention two, were first detected as a result of their physiological activity, not their binding properties.

Even though the physiological response is the single most important criterion for establishing the identity of a specific ligand-receptor interaction, often receptor activity must be studied under conditions where the physiologic response cannot be measured. This is particularly true when trying to detect receptors in fractionated cells or in cells that have been treated with fixatives like formaldehyde or gluteraldehyde. In these situations, the identification of receptor activity has to be based on the properties of ligand binding. These properties must be the same as those established for the intact, responsive cell. Thus, it is not sufficient to measure just ligand-specific displaceable binding (the competition between radiolabeled ligand and excess unlabeled ligand for the receptor). Criteria such as time dependence, ionic and chemical requirements, and cell specificity must also be established [32–34].

Analysis of ligand-receptor interaction is sometimes made difficult by the fact that under physiological conditions binding is not characteristically a bimolecular interaction:

$$(R + L \underset{K_{-1}}{\overset{K_1}{\rightleftarrows}} RL)$$

This is because in the intact cell often the rate of dissociation (K_{-1}) cannot be measured. There may be several different reasons for why a ligand does not dissociate after binding: 1) Ligand binding may be a multivalent step whereby a single ligand can bind to more than one receptor. Galactose terminal glycoprotein [13] and membrane-specific antibodies are examples of these types of ligands. Since the binding of a single ligand to multiple receptors exhibits a rate of dissociation that is a power of the valence of the ligand (that is, $(K_{-1})^n$, where n is the valency and K_{-1} the rate of dissociation), for multivalent ligands the rate of dissociation is so slow that binding is essentially irreversible. 2) Upon binding, ligand may become covalently attached to the receptor, thus precluding a dissociation step. Evidence indicates that some of the cell-bound insulin [18], epidermal growth factor [19–21], and thrombin [20, 22] becomes covalently attached to their receptors. In some cases, covalent binding of ligand to receptor may not be physiological but may represent an artifact of the radioactive labeling technique used to tag the ligand [23]. 3) Once the ligand binds to the receptor, the ligand-receptor complex is internalized. Ligands such as LDL are internalized so rapidly that there is not time for dissociation to take place. This behavior is common for the majority of those receptors listed in Table II, for epidermal growth factor [24], and for human choriogonadotropin [25]. Therefore, in the case of receptors involved in RME, often conventional methods for determining affinity are not applicable. Moreover, it may be impossible to establish whether ligand-receptor interactions exhibit cooperativity or are multivalent. Recently, Wiley and Cunningham [26] devised a particularly useful method for receptor analysis that depends on establishing binding conditions under steady-state conditions rather than at equilibrium. Such a methodology can be used to derive expressions for measuring both the rate of endocytosis and the rate of ligand-induced receptor loss.

B. Physiology

In 1981 Kaplan [27] reviewed the physiologic properties of those cell surface receptors that are involved in specific macromolecule binding. He noted that these receptors could be assigned to one of two catagories: 1) receptors that transport molecules into the cell for further metabolic processing (class II), or 2) receptors that function to directly alter cell behavior or metabolism after ligand binding (class I). Class I receptors can be further divided into those that are able to internalize their ligand (class IA) and those that do not internalize ligand (class IB). Tables I and II list the members of each class along with some of their biochemical properties. The members of each class show similarity in receptor reutilization, surface distribution, ion requirements for ligand binding, and receptor regulation.

C. Reutilization

One of the hallmarks of the class II receptors is that the receptor is reutilized multiple times during its lifetime. Since these receptors function to internalize transport proteins or to clear undesirable macromolecules from the extracellular space, it stands to reason that receptor reutilization would be necessary to maintain a steady-state internalization of the molecules. In many cases, the ligand internalized by these receptors is catabolized in lysosomal compartments; alternatively, the ligand is stored or transported through the cell to the extracellular space. Invariably, the receptor escapes degradation, returns to the cell surface, and mediates the internalization of more ligand (see Receptor Recycling).

In contrast, most class I receptors are not reutilized. Upon binding, the ligand is internalized and the ligand-receptor complex goes to the lysosomal compartment. This process results in an effective reduction of receptors from the cell surface. "Down-regulation" of surface receptors causes the cell to lose its sensitivity to class I molecules. To illustrate, gonadotropin induces steroid synthesis in target cells. However, the gonadotropin effect requires that the ligand and the receptor exist as a complex on the cell surface. Removal of surface receptor-bound ligand by chemical treatment or by internalization

TABLE I. Class I Receptors

Ligand	Function	Regulation of receptor	Surface distribution	Selected references
Class IA Receptors				
Chorio-gonadotropin	Hormone production	Down-regulation	Unknown	25, 35
Epidermal growth factor	Mitosis	Down-regulation	Random or clustered in coated pits	24, 26, 188, 252, 253
Glucagon	Metabolic activity	Down-regulation	Unknown	254
Insulin	Metabolic activity	Down-regulation	Random	255–259
Thyroid-releasing hormone	Hormone secretion or synthesis	Down-regulation	Unknown	260
Somatostatin	Mitosis	Down-regulation	Unknown	261
Growth hormone	Mitosis	Down-regulation	Unknown	36
f-met-leu-phe, polymorpho-nuclear leukocyte	Ion transport, chemotaxis	Down-regulation	Random	39, 262, 263
C_{5a}, polymorpho-nuclear leuko-cyte	Ion transport, chemotaxis	Down-regulation	Unknown	262
Class IB receptors				
Acetylcholine	Ion transport, muscle contraction	Increase in dissociation rate	Clustered	61
Catecholamine	cAMP	Decreased affinity	Unknown	62, 270
IgE, mast cells	Histamine secretion	None	Random	271, 272

results in a reduction in the synthesis of steroid hormone [35]. Down-regulation also affects the sensitivity of cells to a subsequent exposure to the ligand. Studies on the internalization of epidermal growth factor [24, 36], human growth hormones [37], and choriogonadotropin [25] indicate that internalization of ligand is correlated with receptor loss. It is less clear whether insulin can down-regulate its receptor [38]. Perhaps the effect of insulin on receptor reutilization depends on the cell type.

A noted exception among the class I receptors is those for the polypeptide F-met-leu-phe in polymorphonuclear leukocytes. This receptor, which causes

TABLE II. Class II Receptors

Ligand	Function	Regulation of receptor	Surface distribution	Divalent cations required for binding	Low pH sensitivity of binding	Reference
Low-density lipoprotein	Supplies cholesterol	Cholesterol	Coated pits	+	+	29, 58, 125
α_2-Macroglobulin-protease complex	Removes injurious agents	None	Coated pits	+	+	44, 205, 206
N-acetylglucosamine/mannose terminal glycoproteins	Removes injurious agents	None	Coated pits	+	+	44, 264
Galactose terminal glycoproteins	Removes injurious agents	None	Coated pits	+	+	13, 128,232
IgG, oocyte	Fetal immunity	None	Coated pits	+	+	45
Vitellogenin	Source of protein for embryo	None	Unknown	+	+	273
Fibrin	Removes injurious agents	None	Unknown	+	Unknown	265
Cobalamin, cobalamin II	Supplies vitamin B12	None	Microvilli	+	Unknown	266, 267
Mannose-6-phosphate terminal glycoproteins	Delivers lysosomal enzymes to lysosome	None	Coated pits	−	+	9, 46, 50
IgG, small intestine	Newborn immunity	None	Coated pits	−	−	51, 53, 54
Transferrin	Iron uptake	Iron	Random	−	−	56, 268, 269
Semliki forest virus	Virus infection	None	Random	−	−	273

cells to orient in a gradient of the polypeptide, appears to be capable of recycling. In the presence of the polypeptide, the majority of these receptors leave the cell surface within 20 minutes [39]. The receptors remaining on the cell surface are able to continually internalize the polypeptide by a receptor-mediated process. If the polypeptide is then removed from the culture media, the receptors return to the cell surface without a requirement for protein synthesis [40]. At the plateau level there may be a steady-state internalization and return of receptor to the cell surface in the presence of ligand, but after ligand removal all of the internalized receptors return to the cell surface. These cells apparently contain a mechanism for preventing the receptor from getting degraded.

D. Surface Distribution

Where studied, nearly all receptor-mediated processes occur through coated pits (see Endocytosis). Some of these receptors are initially randomly distributed in the membrane and require ligand binding before moving into coated pits. Other receptors appear to be preclustered in coated pits. All of those examples of apparent preclustered receptors belong to the class II family of receptors. Evidence exists that the majority of the receptors for LDL in human fibroblasts [41, 42], asialoglycoprotein in hepatocytes [43], mannose terminal glycoprotein and α_2-macroglobulin protease complex in macrophages [44], IgG in chicken oocytes [45], mannose phosphate terminal glycoprotein in fibroblasts [46], and ferritin in reticulocytes [47] are prelocalized in coated pits. It has not been determined in each case that ligand binding did not induce movement into the coated pit during the period of labeling [see 48, 49]; however, in the case of LDL and IgG this is unlikely. The possible mechanisms of receptor clustering in coated pits are considered later.

E. Ion Requirements for Binding

In contrast to class I receptors, ligand binding to class II receptors is usually sensitive to cation concentrations and pH. Whereas most of the receptors that require calcium for binding also dissociate at low pH, some receptors—for example, those for mannose phosphate terminal glycoproteins [50]—are strictly pH-sensitive. Since class II receptors undergo reutilization during internalization, the sensitivity of the receptor to low pH and/or to calcium may play a role in this process (see Receptor Recycling).

Although most class II receptors exhibit a lower ligand affinity at low pH, the opposite is true for the IgG receptors in neonatal intestinal epithelial cells. The lower the pH, the more tightly this ligand binds to its receptor [51–53]. This seems to be a specialization that permits the receptor to function in the relatively acidic environment of the neonatal intestine. Interestingly, once internalized, the IgG appears to remain bound to its membrane and is transported to the lateral border of the cell. At this site, exocytosis takes place and the IgG dissociates at the higher pH (7.2 to 7.4) of this compartment [54, 55].

A class II receptor that is a total exception to this rule is the receptor for transferrin. Neither calcium nor pH affects the binding of transferrin to its receptor. Once internalized, transferrin is able to release its iron, possibly because of the low pH environment of the endocytic compartment or the lysosome [56]; however, alternative interpretations have been proposed [57]. Even though iron is removed from the transferrin in the cell, the apotransferrin does not accumulate and is not degraded; rather, it is released intact from the cell [56]. Thus, this receptor may function to transport transferrin to a compartment where the iron is deposited and then return during membrane recycling to the cell surface.

The ionic requirements for class II receptor activity seem to be an important aspect of their ability to function as transport receptors: 1) Receptors that accumulate extracellular ligand without any depletion of receptors from the cell surface depend on either divalent cation or neutral pH for ligand binding. 2) Among the receptors involved in transcellular transport, whether it be to another surface of the cell or to the same surface from which it originated, ligand binding is neither low pH-sensitive nor low cation-sensitive. 3) Receptors that function primarily to direct ligands to various intracellular compartments, such as the mannose-6-phosphate receptor, are sensitive to low pH but not to low divalent cation concentration. More work is required to understand how the sensitivity to various ions in the environment regulates receptor activity.

F. Regulation

The mode of regulation of class I receptors is different from that of class II receptors. At least some of the class I receptors (class IA) appear to be regulated by ligand-induced loss of receptors during internalization. Such receptors exhibit a high affinity for their respective ligands; however, it is unclear whether or not the ligand-induced down-regulation is a specific cellular response that is designed to decrease the sensitivity of the cells to the ligand. Class II receptors, in contrast, appear to be regulated by the metabolic products derived from the internalized ligand. (In most cases these receptors are not regulated at all. See Table II.) As an example, the LDL receptor is regulated by the cholesterol liberated from the degradation of LDL in the lysosome [58]. Likewise, iron may be able to regulate the number of transferrin receptors on the cell surface [59]. Iron seems to decrease the rate of transferrin receptor synthesis without altering the rate of receptor degradation.

Some class I receptors (class IB) do not appear to be regulated by either the binding of ligand or the breakdown products of that ligand. A prime example of such a receptor is the acetylcholine receptor [60]. These receptors are asymmetrically distributed on the cell surface and appear to be concentrated at neuromuscular junctions. The binding of acetylcholine changes the ion permeability of the cell, which leads to muscle contraction. Because of low affinity of acetylcholine for its receptor, ligand-receptor dissociation takes place very rapidly. Furthermore, the ectoenzyme acetylcholine esterase hydrolyzes receptor-bound acetylcholine. Therefore, there is not a need to internalize the ligand to terminate the acetylcholine mediated signal [61].

The β-adrenergic receptor is similar to the acetylcholine receptor. With increased time of binding of this ligand, which causes an increase in intracellular cyclic AMP concentration, there is a decrease in the rate of cyclic AMP production. Part of the decreased response may be due to receptor internalization [62]. However, much of the lost activity appears to be due to receptor occupancy. Isolated membranes exposed to the ligand also exhibit a time-dependent decrease in sensitivity to adenyl cyclase induction. This loss of activity may be due to a decrease in ligand-receptor affinity; in other words, the binding of ligand induces a conformational change in the receptor such that it can no longer bind ligand. GTP may be involved in modulating binding affinity [63]. In this case, modulation of receptor activity does not require ligand internalization but is due to the special relationship of this receptor to other regulatory mechanisms in the cell.

Since the class IB receptors have other mechanisms for modulating their activity, it seems likely that class IA receptors must be internalized to regulate the cellular response to ligand. Much more research is needed to establish this point. Nonetheless, a comparison of class I and class II receptors illustrates that the cell utilizes a general endocytic process to achieve quite different physiologic events.

III. ENDOCYTOSIS

Despite the large number of receptors that mediate the internalization of extracellular molecules, the process of endocytosis always involves the conversion of a planar segment of membrane into an indentation or pit. Subsequently the opposing portions of membrane fuse at the lateral margins of the indentation to form an endocytic vesicle [64]. The size of the invagination and the vesicle formed varies from several micrometers in diameter to less than 0.1 μm. Furthermore, different regions of surface membrane may participate in this process.

In 1931, Lewis [65] first described the process of fluid-phase endocytosis, which he termed pinocytosis. Using phase-contrast light microscopic techniques, he was able to visualize in cultured macrophages 1–2 μm in diameter vacuoles that formed in areas of the cell that were active in membrane ruffling.

The large vacuoles that were formed by this process entered the cell and eventually diminished in size. Many other investigators have confirmed Lewis' observations in a variety of different cultured cells [66, 67]. With the advent of the electron microscope, Palade [68] and Bennett [69] described vesiculations of the plasma membrane that ranged in size from 0.05 μm to 0.2 μm in diameter, which they interpreted to be invaginations of the cell surface that were involved in endocytosis. These investigators were able to visualize a form of endocytosis that could not be detected with the light microscope.

These light- and electron-microscopic studies have led to the concept that the endocytosis of molecules can be divided into two general catagories—macropinocytosis and micropinocytosis [66, 70, 71]. The early electron microscopic studies suggested that micropinocytosis could be further subdivided. Palade [72, 73] first described flask-shaped invaginations of the endothelial cell plasma membrane that had a circular body 60–70 nm in diameter with a narrow neck that opened to the extracellular space [reviewed in 64, 71]. Subsequent tracer studies implicated these structures as sites of endocytosis [71]; however, more recently the idea that these structures mediate endocytosis has been challenged [71, 74–76]. In 1964, Rosenbluth and Wissig [77] described the endocytosis of ferritin in toad spinal ganglia cells, a process that involved "coated" vesicles. In the same year, Roth and Porter [78] reported that similar membrane specializations were increased in oocytes that were internalizing yolk protein. Thus, micropinocytosis might occur through either flask-shaped vesicles or coated vesicles.

Where examined with appropriate visual markers, almost all of the receptors involved in RME have been shown to internalize ligands through coated regions of membrane [7, 8]. In this review we will therefore confine our analysis to the function of coated pits and coated vesicles as vehicles for RME.

The terms "coated pit" and "coated vesicle" were first used by Rosenbluth and Wissig [77], to describe surface invaginations and vesicles that seemed to be involved in the uptake of ferritin molecules in nervous tissue. Such structures have been noted by other investigators and termed "bristle-coated structures " [78], "rhopheosomes" or "rhopheocytic vesicles" [79, 80], "complex vesicles" [81], "alveolate vesicles" [82], "acanthosomes," and "amorphous-coated structures" [83, 84]. In all cases, the names of these structures were derived from the fact that they characteristically are decorated by a dense cytoplasmic coat that often appears in the form of projections from the cytoplasmic face of the plasma membrane.

First by thin-section electron microscopy [85, 86], and then by rapid freeze-etch microscopy [87], it has been determined that the coat material consists of an arrangement of pentagons and hexagons joined together to form a polygonal network (see Fig. 1). Kanaseki and Kadota [85] were the

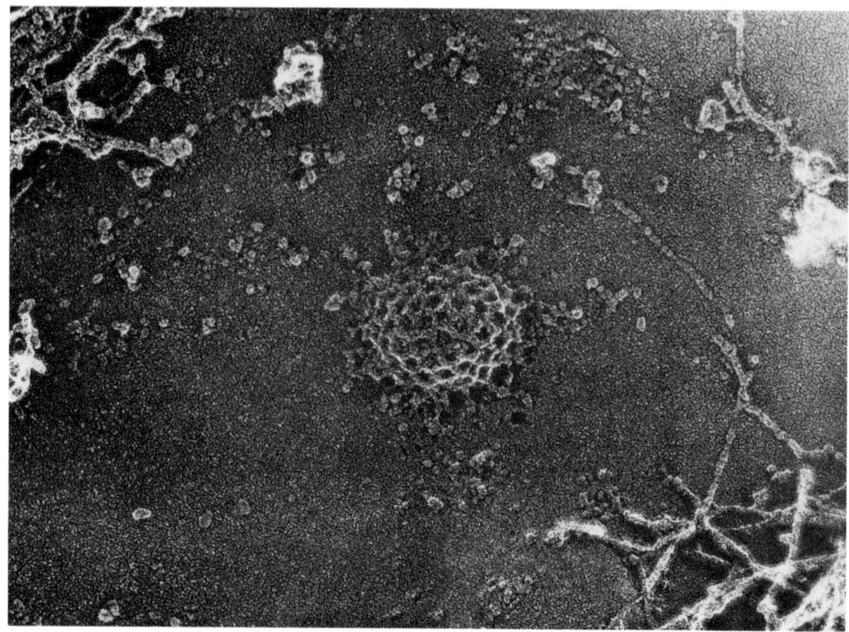

Fig. 1.   Rapid freeze-etch image of the inner membrane surface of cultured fibroblasts. Coated membrane is easily distinguished by the polygonal lattice of the coat material. Each lattice is composed of both hexagons and pentagons. (Courtesy of J. Heuser). × ∼ 148,000.

first to partially purify coated vesicles from nervous tissue; however, Pearse [88] improved this procedure, which made it possible to obtain large numbers of coated vesicles for biochemical and structural analysis [89–100]. Pearse's original electrophoretic analysis of purified coated vesicles suggested that there was one major protein, M_r 180,000, associated with this structure [88, 89]. She named this protein "clathrin" [89]. Other investigators have analyzed the protein composition of coated vesicles in more detail. In addition to clathrin, coated vesicles often contain proteins of M_r 110,000, 55,000, and 33,000–36,000 [94–97].

Several investigators have utilized the coated-vesicle isolation technique to prepare antibodies against clathrin [101–107]. As seen by the rapid freeze-etch techniques in Figure 2, these antibodies specifically bind to the characteristic polygonal lattice of the coated pit. Anticlathrin antibodies have been useful as immunocytochemical probes. In tissue culture cells, anticlathrin has been used to show that coated pits are uniformly distributed on the cell surface [103, 104, 107]; in addition, large pools of clathrin, possibly in

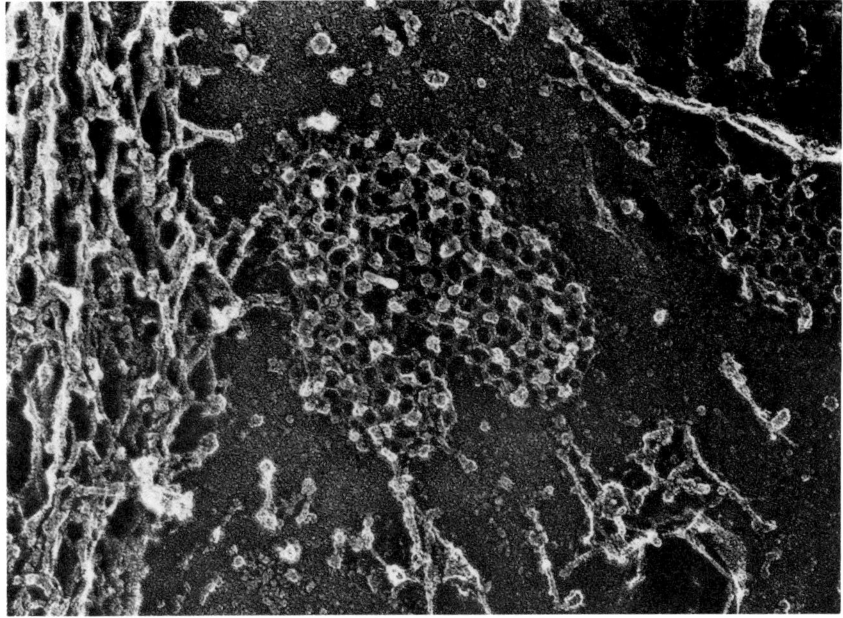

Fig. 2. Rapid freeze-etch image of coated membrane that has been exposed to 0.5 mg/ml anticlathrin antibody [104]. The IgG molecules appear as small globules that decorate the coat protein lattice. (Courtesy of J. Heuser.) $\times \sim 132,000$.

the form of small coated vesicles, are located in the perinuclear area of the cell. These antibodies have also been used to detect specialized arrangements of clathrin in epithelial cells [106] and nerve tissue [102].

Isolated coated vesicles can be disassembled and reassembled in vitro [95–97, 108, 109]. Treatment of coated vesicles with either 2 M urea [95, 97] or 0.75 M Tris [96, 100, 108, 109] solubilizes the clathrin coat. Upon removal of these agents, basket or cagelike structures, which display the normal polygonal morphology, reform. Clathrin cages disassembled in Tris yield a soluble 8.6 S species of clathrin that by negative staining appears as a threefold symmetric structure that is shaped like a pinwheel (Fig. 3A). This subunit, termed a "triskelion" [108], is composed of equimolar amounts of clathrin (M_r 180,000) and one of the two low molecular weight proteins (M_r 36,000 or 33,000). Cross-linking experiments suggest that each M_r 180,000 protein is associated with one low molecular weight protein. Not only are triskelions capable of assembling into polygonal baskets, they also are able to bind to a protease-sensitive site that is present on membrane vesicles

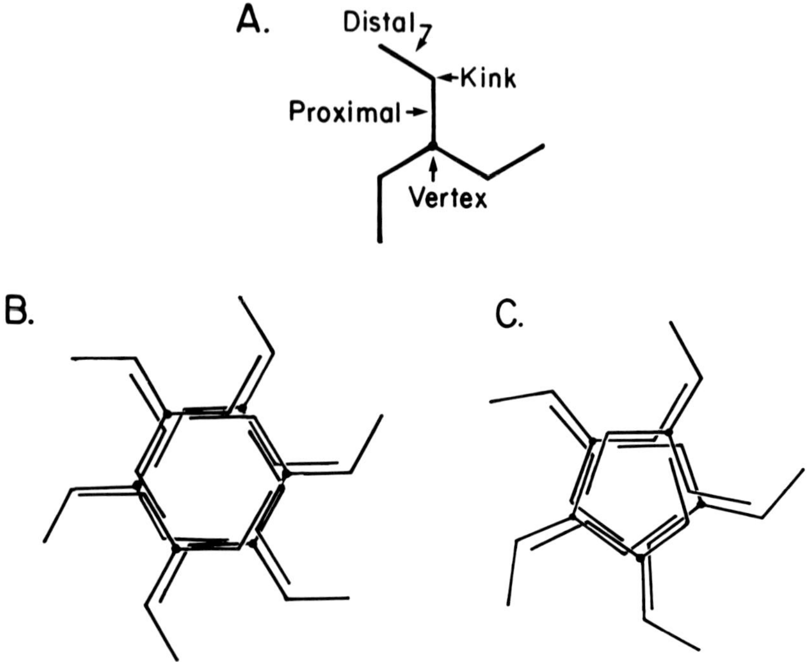

Fig. 3. Diagram illustrating the structure of a triskelion (A) and the various ways that triskelions are arranged to form hexagons (B) and pentagons (C) in the coat lattice. (Redrawn from Crowther RA, Pearse BMF: J Cell Biol 91:790–797, 1981.)

isolated from coated vesicle preparations [110]. The triskelion appears to be the basic subunit of the coated vesicle.

The packing of 8.6 S clathrin into the hexagon and pentagon coats of coated vesicles has been studied by Crowther and Pearse [111]. Negatively stained images of partially reassembled cages show an overlapping arrangement of the triskelion legs to form the sides of each polygon (Fig. 3B, C). The vertex of each triskelion is assumed to be at the vertex of each polygon. Based on this electron microscopic analysis, several packing arrangements have been proposed.

The lipid composition of coated membrane has also been investigated. Isolated brain coated vesicles contain 43% phosphatidyl choline, 30% phosphatidyl ethanolamine, 4% phosphatidyl serine, 12% sphingomyelin, 11% phosphatidyl inositol, and no lysolecithin [88]. In addition, they contain a cholesterol to phospholipid ratio of 0.1–0.3 [89], a value that is lower than that found for most plasma membranes (0.3–1.2 [112]). The cholesterol composition of coated membrane has also been analyzed by electron-microscopic cytochemistry, using the cholesterol-binding agents digitonin [113] and filipin [114]. In contrast to the surrounding plasma membrane, coated regions do not contain the characteristic precipitates formed by these agents when they interact with cholesterol. There is good evidence that these precipitates form only in membranes that contain cholesterol [113, 115]; however, it has not been determined that the presence of membrane cholesterol is sufficient for precipitate formation. Despite the fact that these electron-microscopic studies have been interpreted to mean that coated membrane contains little or no cholesterol, it is also possible that cholesterol is present but that the cytoplasmic coat or some other property of coated membrane prevents the formation of filipin-cholesterol precipitates.

Besides the clathrin coat and the possible low cholesterol composition, coated surface membrane also has a unique population of intramembrane particles. Orci et al [116] determined that fibroblast-coated pits contain twice as many intramembrane particles per unit area as the surrounding membrane. Moreover, these particles are ~20% larger than the particles in the uncoated membrane. Presumably, these particles reflect the presence of unique membrane proteins or protein-lipid aggregates. To date, their identity has not been determined.

Coated pits and coated vesicles are found associated with two major membrane compartments: cell surface membrane, and Golgi apparatus membrane. Their function at these two surfaces appears to be to move molecules from one region of the cell to another. Coated membrane has been observed to be associated with 1) the exocytosis of secretory products from the Golgi [117], 2) transport of lysosomal enzymes from the Golgi [118], 3) the endocytosis of extracellular macromolecules [7], 4) membrane recycling [119], and 5) transport of tyrosinase [120]. Coupled with the activity of numerous types of receptors, coated membrane appears to be designed to carry out vesicular transport.

Nearly all of the investigators who have studied the endocytosis of tracers through coated pits report that the label is sometimes seen in coated vesicles. It has therefore been assumed that coated pits transform into coated vesicles during endocytosis. Recently Willingham, Pastan, and co-workers have proposed an alternative route of internalization: Coated pits concentrate receptors with attached ligand, but internalization occurs through uncoated membrane that buds laterally from the coated pit [101, 121–123]. These investigators have presented three lines of evidence that supports this model. 1) All of the coated vesicle profiles seen in cells fixed after incubation at 4°C are permeable to the cationic dye, ruthenium red. Therefore, all apparent coated vesicles must actually be coated pits that are sectioned in a plane that does not show continuity with the extracellular space [124]. 2) Immunocytochemical labeling of fibroblasts with anticlathrin antibody does not reveal a soluble pool of clathrin [101]. Such a pool should exist if coated pits become coated vesicles because the coat protein would have to dissociate from the coated vesicle and return to the cell surface. 3) Microinjected anticlathrin antibodies do not inhibit endocytosis of α_2-macroglobulin [123]. Inhibition would be expected during coated pit–coated vesicle–coated pit recycling because the antibody should aggregate dissociated clathrin and prevent it from returning to the cell surface.

Despite these provocative findings, other evidence suggests that coated pits do transform into coated vesicles during endocytosis. In cells that are precisely oriented so that the surface membrane is perpendicular to the plane of section, it is possible, using tracer techniques, to identify all of the intermediates in the conversion of a coated pit to a coated vesicle [125]. Furthermore, in some cells it is possible to find coated-vesicle profiles labeled with exogeneous tracer that are deep within the cytoplasm of the cell [118, 126, 127]. Most likely such structures are not connected to the cell surface. Other studies with ruthenium red have identified coated vesicle intermediates during endocytosis. Wall et al [128] found that at least 50% of the coated-vesicle profiles in hepatocytes were impermeable to this dye. These hepatocyes were fixed at 37°C. Maybe at 4°C coated vesicles do not exist because endocytosis is inhibited. Even when fibroblasts are fixed at 37°C, many of the coated-vesicle profiles are not permeable to ruthenium red [124]. Finally, the ease with which coated vesicles can be isolated, sometimes containing either receptors involved in receptor-mediated endocytosis [129, 130] or endocytosed ligand [130, 131], strongly suggests that they exist in cells.

In our opinion coated vesicles do form during endocytosis, and this vesicle is an important but short-lived intermediate in the endocytic pathway. At 4°C, coated-pit formation takes place, but coated vesicles are not able to form. In future studies, serial sectioning techniques should be used to determine whether coated vesicles are actually coated pits.

A. Mechanisms

Once a ligand is bound to its receptor and the ligand-receptor complex is located over a portion of plasma membrane that is undergoing endocytosis, the receptor with its attached ligand is passively internalized. Although certain polyvalent ligands may induce curvature in membranes in a manner that leads to endocytosis [132], most likely it is the intrinsic properties of the membrane in this region that are responsible for the endocytic event. Regardless of whether endocytosis occurs by smooth-surface membranes or coated-surface membranes, the event requires that a planar surface of membrane become indented or curved. Curvature of this portion of membrane then increases to the point where opposing straight segments of membrane fuse. Despite a wealth of morphologic detail, there is virtually nothing known about the molecular mechanism underlying these membrane shape changes. However, there is enough information to piece together a model of how endocytosis might take place. This model may have some heuristic value for directing future research on endocytosis. [This subject has been briefly treated in 64, 133.]

A reservoir of information that can be used to construct such a model has been provided by those interested in the maintenance of erythrocyte biconcave morphology. In 1968, Deuticke [134] showed that erythrocytes undergo characteristic shape changes in response to the presence of amphipathic molecules. Anionic or nonionized molecules such as fatty acids, saponins, or barbiturates cause erythrocytes to become crenated, a shape that is characterized by numerous membrane projections or spines (Fig. 4A). When erythrocytes are exposed to cationic substances such as local anesthetics, antihistamines, and primaquine, the erythrocytes become cup-shaped (Fig. 4C), an extreme form of the biconcave morphology (Fig. 4B). These agents were found to be mutually antagonistic so that cup-forming agents could reverse the effects of crenation-causing agents. Deuticke concluded from these studies that these agents differentially affected either the outer bilayer (to cause crenation) or the inner bilayer (to cause cupping) of the erythrocyte membrane.

Other investigators have confirmed and extended these studies [135–142]. Sheetz and Singer [136, 137] concluded that crenators (anionic amphipathic molecules) selectively intercalate into the outer half of the bimolecular lipid leaflet and that cup formers (cationic, amphipathic molecules) incorporate into the inner leaflet of the lipid bilayer. Studies of shape change in erythrocyte ghosts revealed that there are other ways to modulate membrane shape [138, 139, 141, 142]. ATP-depleted erythrocyte ghosts are normally crenated; however, upon exposure to ATP, the ghosts become cup-shaped. The ATP-induced shape changes are accompanied by an increase in the phosphorylation of spectrin—an inner surface membrane protein that is associated with the transmembrane protein, glycophorin. Since these ATP-mediated shape changes are inhibited by globular actin, it may be that actin-spectrin interactions are involved in this process. This conclusion was supported by the observation that the cross-linking of spectrin by antispectrin antibodies promoted similar shape changes [138].

Because of these studies, Sheetz and Singer [136] proposed the bilayer coupled hypothesis to explain erythrocyte membrane shape changes. This hypothesis states that the two layers of lipid in the membrane are coupled so that expansion of the outer layer will cause a linear segment of membrane to evaginate, but the expansion of the inner layer of this membrane will result in invagination. Three separate mechanisms can cause the differential expansion of the membrane: 1) preferential intercalation of molecules into

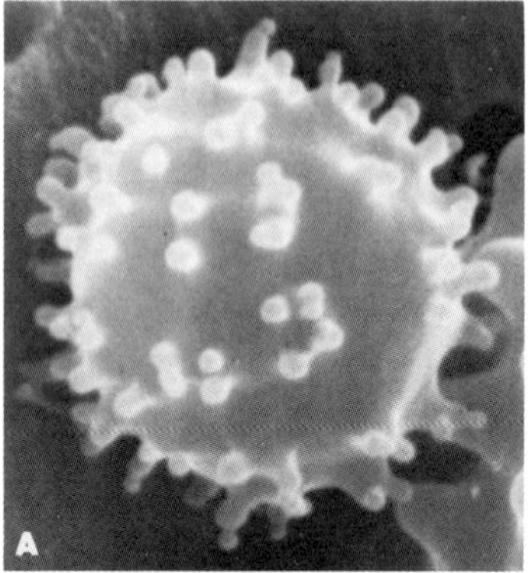
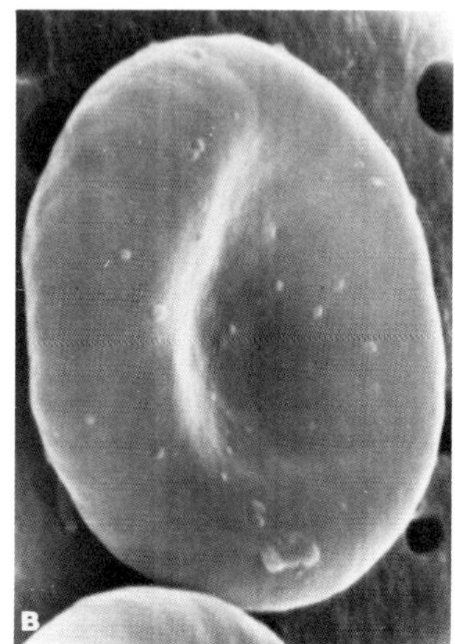
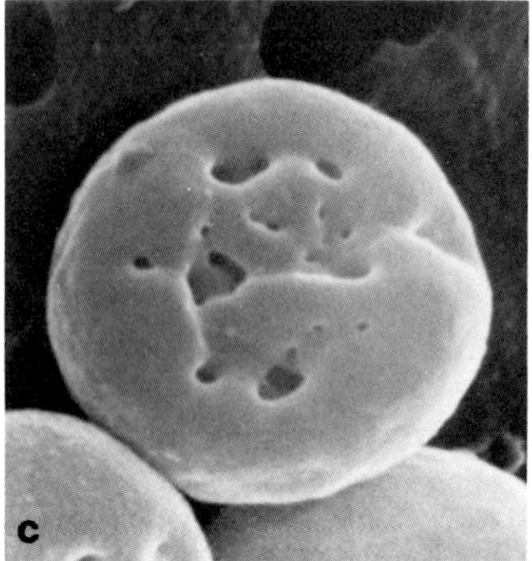

Fig. 4. Scanning electron micrographs of crenated red cells (A), normal biconcave-shaped red cells (B), and cupped red cells (C). (Courtesy of M. Sheetz.)

one of the two lipid layers; 2) a change in the charge of the lipid molecules in one layer relative to the other; or 3) aggregation of proteins in one layer relative to the other. In 1974, Evans [143] stated it another way. Any agent that either reduces the interfacial free energy of the outer layer of a membrane or increases the interfacial free energy of the inner layer will cause erythrocytes to become crenated. An opposite differential change in the interfacial free energy will cause the cup-shaped transformation.

The process of erythrocyte cupping, which can be considered to be analogous to membrane invagination during endocytosis, has been mathematically analyzed by Brailsford et al [144]. They concluded that the bilayer-coupled hypothesis does not sufficiently explain the cupping process. To account mathematically for the measured shape changes during cupping, in addition to the induction of some intrinsic curvature, there must also be a phase change in either the hydrophobic interior or in the bimolecular lipid leaflet such that this region of membrane becomes more fluid. The increased fluidity is required for the two halves of the bilayer to slide with respect to each other. This in turn is necessary for the cup shape to be a stable energy state of the membrane. Membrane intercalating agents would function to expand the bilayer and decrease the free energy at the inner membrane surface. The conclusions reached by these investigators are in agreement with Evans [143]; furthermore, they defined the fluidity of the membrane as an important regulator of cupping.

Even though there are risks associated with extrapolating data from one cell type to another, the analysis of cup formation in erythrocytes provides a convenient reference point for constructing a molecular model of endocytosis. Obviously, cupping involves macroinvaginations of surface membrane whereas endocytosis only requires microinvagination. But the conclusions from the cupping studies do not seem to depend on the amount of membrane area participating in the shape change. Furthermore, there is evidence that cup-forming agents such as ATP [145–147] and certain cationic amphipathic molecules [148, 149] can induce endocytosis in erythrocytes or erythrocyte ghosts.

A model of endocytosis begins with the assumption that membranes consist of many different types of lipid and protein molecules, which are organized into a mosaic of low-fluidity and high-fluidity domains (Fig. 5A). The two lipid layers are able to slide with respect to each other in the more fluid areas, but at the margin of these areas movement is retarded. Specific cytoplasmic factors interact selectively with the inner layer of the fluid regions and impart an intrinsic curvature to this segment of membrane (Fig. 5B). Curvature could be created by expanding the inner bilayer or by physically deforming the membrane. With progressive invagination, the lateral edges of the less fluid areas are drawn together (Fig. 5C). These edges eventually fuse to form the endocytic vesicle (Fig. 5D). Therefore, the formation of endocytic vesicles depends on 1) the existence of small areas of fluid membrane integrated among areas of less fluid membrane; 2) the existence of those factors that induce membrane curvature; and 3) the ability of these factors to interact with specific segments of membrane. The model specifies the minimum structural requirements but not the order of their interaction. It may be that those agents that induce curvature also increase the fluidity of the presumptive endocytic membrane.

This model predicts that endocytosis is an organized cellular process that depends on specific molecular arrangements in the plasma membrane. Whereas coated regions of membrane appear to fulfill this prediction, not enough is known about the chemistry of those smooth-surfaced membranes involved

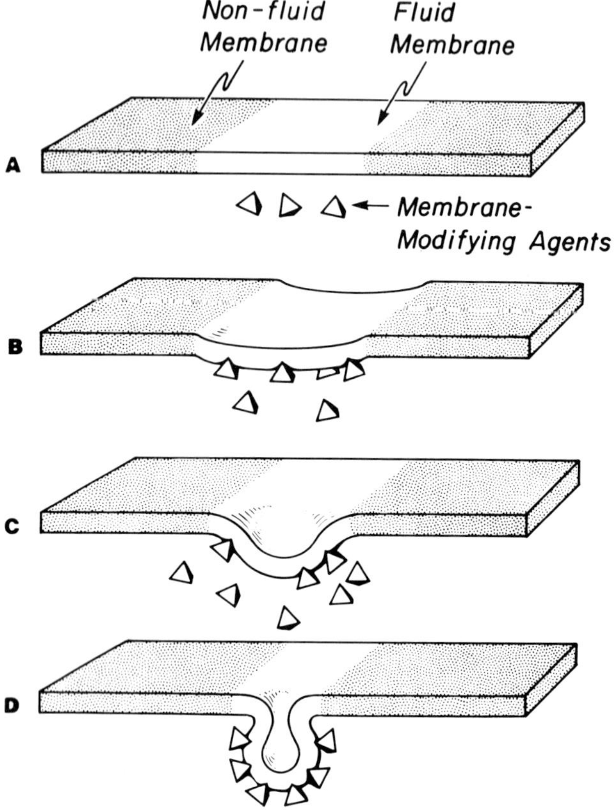

Fig. 5.   In diagrammatic form, a molecular model of endocytosis. (See text for description.)

in endocytosis to analyze this process. Since RME occurs through coated membrane, we will compare what is known about this process with the proposed model.

The polygonal arrangement of triskelions that line the cytoplasmic surface of the coated membrane is a good candidate for an agent that induces the intrinsic curvature required for endocytosis. Exactly how this protein complex would work is not clear. Possibly it simply clusters transmembrane proteins, such as those seen in freeze-fracture images, and these clusters alter membrane curvature. However, a more intriguing idea, first suggested by Kanaseki and Kadota [85], is that the polygonal lattice undergoes a rearrangement, which causes the transformation of the planar segment of membrane into a curved segment. Heuser's [87] study of rapidly frozen membrane has more precisely defined the lattice rearrangement. In planar coated membrane, there is a predominance of hexagons relative to pentagons in the lattice. However, as the membrane becomes more indented, there is an increase in the number of pentagons relative to hexagons. As discussed by Thompson [150], although a lattice of pure hexagons can be planar or curved, it is mathematically impossible to form a closed structure from such a lattice. Only by introducing perturbations into the lattice—for example, by inserting pentagons—can a closed structure be formed. Thus, the increase in the number of pentagons that accompanies the change in curvature of coated membrane during invagination is undoubtedly necessary to form a closed lattice around the coated vesicle. Heuser also found that often pentagon-heptagon translocations were present in the curved lattices. He suggested that these may be intermediates in the lattice rearrangement. Yet to be determined is whether this lattice rearrangement drives the endocytic process or instead is a passive response to an increase in membrane curvature that is generated by other mechanisms. However, one cannot help being impressed that the clathrin lattice has many of the attributes of a structure that could induce the membrane curvature necessary for endocytosis.

A second basic tenet of the proposed model is that membrane at sites of endocytosis is more fluid than the surrounding membrane. If this is true, then a decrease in fluidity of these areas would retard or inhibit the endocytic process. The most effective way to inhibit coated-vesicle-mediated endocytosis is to lower the temperature to 4°C. One of the many effects of lowering the temperature is to decrease the fluidity of the membrane as a result of phase changes in the lipid bilayer [151–153]. More significant, the fatty acid composition of the phospholipid bilayer in macrophage membranes has been shown to effect endocytosis [154, 155]. When mouse peritoneal macrophages are incubated in serum-free medium that contains either a 19:0 or trans-18:1 fatty acid complexed with bovine serum albumin, within 8 hours 25% of the phospholipid fatty acid in the membrane is derived from this exogenous source. The incorporation of either of these fatty acids into the membrane results in an increase in the saturated to unsaturated fatty acid ratio in the phospholipid bilayer. This change is correlated with a 40% decrease in fluid-phase endocytic activity [154] and an increase in the rigidity (a decreased fluidity) of the plasma membrane [155]. Although it was not determined in these studies that bulk-phase endocytosis was occurring through coated pits and coated vesicles, the activation energy of 17–25 Kcal $\times$ deg^{-1} $\times$ M^{-1} for this process is the same as the activation energy for the endocytosis of asialofetuin [156] and α_2-macroglobulin-protease complex [157], proteins that are known to enter cells through coated pits and coated vesicles [44, 128]. Therefore, artifically decreasing the fluidity of the plasma membrane seems to have a marked effect on the ability of the cell to endocytose extracellular molecules, possibly via coated pits and coated vesicles. Finally, there is the evidence that coated regions of membrane may have a lower cholesterol content. Depending on the phospholipid environment, cholesterol can decrease membrane fluidity at physiological temperatures [151–153]; therefore, an absence of cholesterol in the coated regions of membrane could mean that it is a region of relatively higher fluidity than the surrounding membrane.

From the above analysis, we suggest that coated pits and coated vesicles have some of the membrane properties predicted to be required for a segment of membrane that can undergo endocytosis. The model and supporting data indicate new directions of research into the mechanisms of coated-pit–coated-vesicle–mediated endocytosis. Among the most intriguing questions is how the coated region becomes differentiated with respect to lipid composition, intramembrane particle composition, and coat deposition. These regions of membrane are dynamic in that they are continually budding off to form endocytic vesicles and re-forming [158]. Marsh and Helenius [159] have estimated that 1,500–3,000 coated vesicles can form per minute per cell; recycling of these regions of membrane must be very fast! It will first be necessary to determine whether or not the coat protein that is arranged into the polygonal lattice of the coat is broken down into subunits during this process or whether the lattice dissociates intact and then reassociates with surface membrane. Although Willingham et al [101] did not detect a cytoplasmic clathrin pool in fibroblasts, Cheng et al [102] did find free clathrin in the cytoplasm of nerve tissue. Furthermore, Heuser's studies suggest that the clathrin lattice is formed from subunits that are deposited on the membrane [87]. Unanue et al [110] recently found that isolated triskelion subunits are able to bind to isolated coated vesicles that have been treated to remove the clathrin lattice. The binding site may be a protease sensitive M_r 110,000 protein, which suggests that triskelions recognize a specific type of protein. The identity of this protein remains to be established. Future research efforts should focus on these types of observations to achieve a better understanding of the molecular mechanisms involved in endocytosis.

Since coated surface membrane has the properties of a membrane specialized for endocytosis, cells could utilize the coated-pit–coated-vesicle system to regulate both the temporal and the spatial arrangement of endocytosis. Several studies provide evidence that under certain conditions endocytosis through coated membrane can be stimulated, suppressed, or spatially rearranged on the cell surface.

1. Stimulation. In many cultured cells, the number of coated pits is fairly constant—about 2% of the cell surface [41, 116, 160]. However, there are a number of examples where cells are able to increase the number of surface coated regions in response to specific stimuli. Roth and Porter [78] were able to elucidate the role of coated pits in adsorptive endocytosis by studying the mosquito oocyte because these cells accumulate massive numbers of coated pits in response to a blood meal. More recently, Fisher and Rebhun [161] have discovered that sea urchin egg membrane accumulates coated pits following fertilization, and that these pits transform into coated vesicles that take up extracellular markers such as horseradish peroxidase. Some cells accumulate coated pits (as much as a threefold increase) when exposed to either nerve growth factor [162] or epidermal growth factor [163]. Finally, following nerve stimulation, there is an increase in the number of coated pits and coated vesicles found at the presynaptic terminals [119]. In most cases, the appearance of coated pits in these cells is transient, lasting only several minutes, and they seem to be associated with membrane retrieval [119, 161, 162].

2. Suppression. There is also evidence that coated-pit-mediated internalization can be suppressed. For example, during mitosis, the endocytosis of concanavalin A (Con A) is completely inhibited [164]. Since this lectin is known to be internalized by coated membrane [165], most likely coated-pit–coated-vesicle function is inhibited during cell division. Possibly this is achieved by calcium-regulatory proteins. Salisbury et al [166] have reported that the calmodulin inhibitor trifluoperazine dihydrochloride (TFP) can reduce the number of coated pits recruited to the cell surface in response to the binding of anti-IgM in lymphoblastoid cells. TFP-treated cells appear not to internalize as much anti-IgM [126]. By both biochemical [166, 167] and immunocytochemical light microscopy [168], calmodulin has been found to be associated with isolated coated vesicles and regions of Con A endocytosis. Unfortunately, no one has shown by electron-microscopic immunocytochemistry that coated pits contain this Ca^{++}-regulatory protein. Nevertheless, the observation that the activity of coated membrane could be regulated by calmodulin is very intriguing.

3. Spatial arrangement. In addition to the rate of endocytosis by coated membrane, the distribution of coated pits on the cell surface is sometimes modulated. A macrophage cell line, J774, when treated with colchicine, develops a bulge or protuberance that is separated from the cell body by a constricted neck region [169]. Such cells display a 50% increase in coated-pit density that is correlated with a reduced rate of endocytosis [170]. Furthermore, the protuberance contains significantly more coated pits than the body, which suggests that a rearrangement in the coated-pit distribution took place. A related observation is that in several different cells, the capping of surface determinants results in a preferential arrangement of coated pits at the cap and these then participate in removing the capped ligand from the cell surface [165, 166, 171]. Possibly the rearrangement of coated-pit distribution is associated with cytoskeletal-membrane interactions. Salisbury et al [166] presented evidence that actin can interact with coated pits and coated vesicles in capped regions of the cell during endocytosis. Furthermore, in vitro coated vesicles may bind to actin [97]. However, no one has shown that this is a specific interaction. It is possible that the high concentration of actin in the capped region adheres nonspecifically to coated vesicles. Finally, Sattilaro et al [172] have reported that coated vesicles can interact with microtubules in vitro via microtubule-associated proteins.

B. Receptors and Coated Membrane

1. Efficient internalization. To date 22 macromolecules or macromolecular complexes such as viruses have been shown to enter cells by RME [reviewed in 7, 8]. Of these, 12 have been shown by morphologic techniques to enter the cell through coated pits. (It is also possible that c3b receptors function through coated pits [173].) The site of entry for the remaining have yet to be identified; most likely, however, all these molecules enter through coated pits because each of them displays approximately the same kinetics of internalization as the molecules that are known to enter through this route. Once bound to a receptor, most of these molecules leave the cell surface within 12–15 minutes. Goldstein et al [7] have proposed that the rapid kinetics of internalization is a function of the ability of the receptor to interact with coated pits. If this is the case, then 1) receptors that mediate the internalization of ligands must be able to interact specifically with coated pits; and 2) receptors that do not internalize their ligand should have properties that prevent them from interacting with coated pits. Not only do the coated pits constitute a mechanism for efficiently internalizing molecules, but, because they are a primary site of endocytosis, exclusion of determinants from the coated pits is necessary for those ligand-receptor complexes that have a long lifetime on the cell surface.

The relationship of efficient internalization to coated membrane function has been clarified by studying LDL receptor activity in two unique cells: 1) the internalization defective, human fibroblast [174–176], and 2) the human carcinoma A-431 cell [160]. The efficiency of LDL internalization can be determined by measuring the ratio of internalized and degraded LDL to surface-bound LDL in a unit period of time. This fraction, called the "internalization index" [160, 174], averages ~30 in normal human fibroblasts, which have 50–70% of the LDL receptors in coated pits [125]. In contrast, internalization-defective cells have an internalization index of ~5 [174], and only 3% of the LDL receptors are in coated pits [176]. Similar results were found for the A-431 cells. This cell displays over six times more LDL receptors on its cell surface than the normal human fibroblast, but only 4% of these receptors are in coated pits [160]. The internalization index for the A-431 cell is ~6. Therefore, in two different cell types, the efficiency of internalization of LDL was found to be directly proportional to the number of surface receptors associated with coated pits. These studies emphasize that receptors that cannot associate with coated pits internalize their ligand very slowly if at all.

Perhaps a better way to assess the contribution of coated membrane is to study membrane determinants that do not function through coated pits. Such a determinant is β_2-microglobulin, which is found on the surface of many cells. When human fibroblasts are incubated with anti-β_2-microglobulin, numerous small clusters of antibody form, which are randomly distributed on the cell surface [177]. Following the addition of an antibody against the bound IgG, these clusters become linearly arranged on the cell surface. Furthermore, some of these clusters are internalized through noncoated membrane [178]. We have compared the relative rate of internalization of antibodies against β_2-microglobulin, which are not internalized by coated pits, with antibodies against the LDL receptor, which are internalized through coated pits [42, 179] (Fig. 6). Cells were incubated at 37°C with either anti-LDL receptor (Fig. 6A, B) or anti-β_2-microglobulin (Fig. 6C, D) and then fixed. One set of cells was treated directly with FITC-conjugated antirabbit IgG (Fig. 6A, C), whereas the other set was first treated with Triton X-100 to premeabilize the cell. After 10 min all of the anti-LDL receptors had left the cell surface (Fig. 6A), but after 15 min uniform surface staining was seen in the anti-β_2-microglobulin-treated cell (Fig. 6C). All of the anti-LDL receptors had entered the cell (Fig. 6B), but little intracellular anti-β_2-microglobulin was detected (Fig. 6D). Even when β_2-microglobulin is clustered by the addition of a second antibody (anti-anti-β_2-microglobulin), the complex remains on the cell surface for over 30 min [see Fig. 2, Ref. 178]. This experiment demonstrates visually the effectiveness of coated membrane as a mechanism for internalizing surface-bound determinants.

Thus, surface membranes contain molecules that can associate with coated pits (eg, receptors involved in RME) or are excluded from this region of membrane (eg, θ and H63 antigens [180]). One of the largely unanswered questions in receptor biology is what determines how these components of the membrane interact.

2. Receptors in coated pits. Presumably the molecular properties of both the coated pit and the membrane determinants specify whether the two can form a stable association. Receptors involved in RME may be able to bind specific extracellular ligands and at the same time bind to coated pits. For example, the mutant human fibroblast that is able to bind but not internalize LDL has receptors that are unable to associate with coated pits [176]. This mutation appears to be allelic for the LDL receptor [174]. Therefore, the receptor must have the proper three-dimensional structure to get into coated pits. The clustering of receptors into this region of membrane may involve

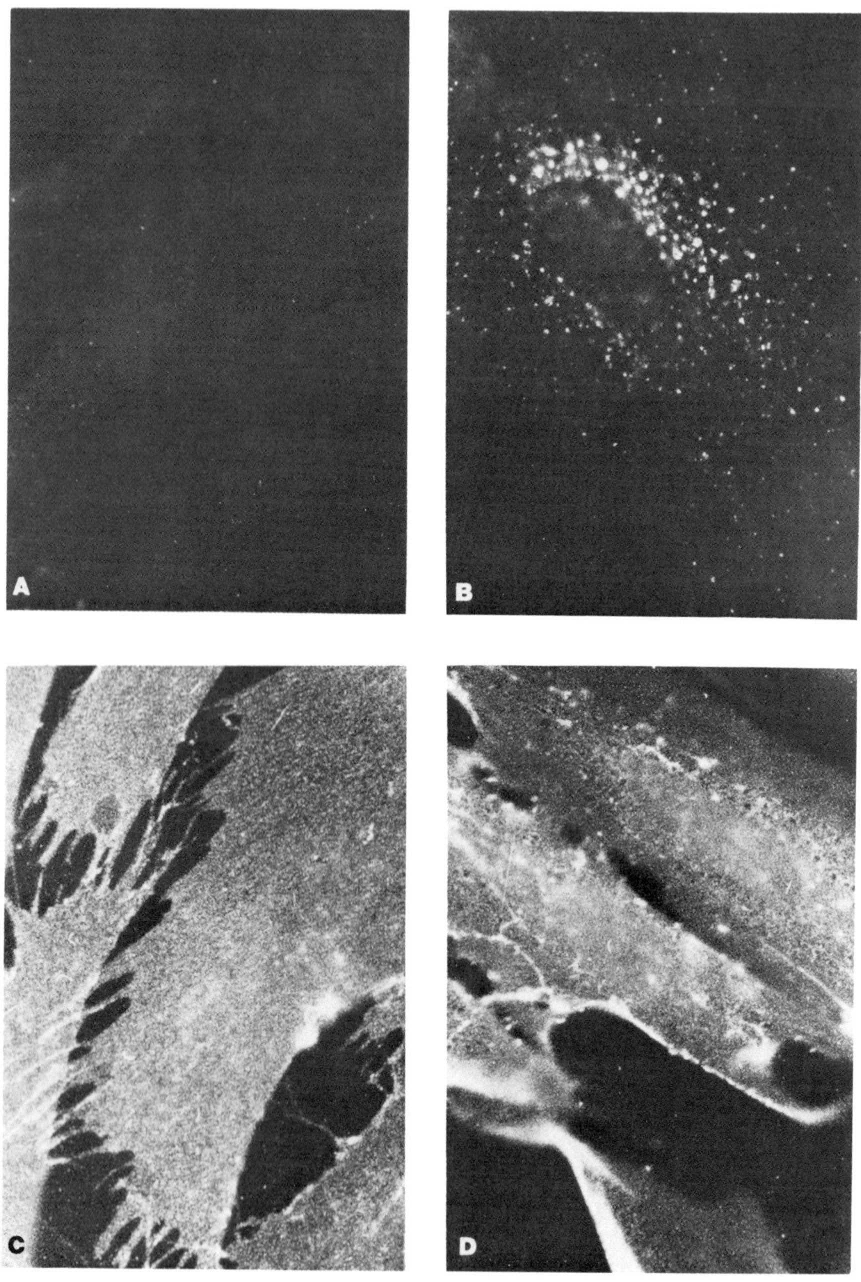

some type of capture process. Goldstein et al [181] have recently presented a mathematical model of this process. If LDL receptors are irreversibly captured by coated pits and the diffusion constant for the receptor is 10^{-11} cm^2/sec [182], then there is enough time during recycling for randomly inserted receptors to carry out coated-pit-mediated internalization of LDL. The coated-pit capture event would be a function of receptor-coated-pit recognition.

An alternative idea is that receptors are passively carried into coated pits. Bretscher [183] has suggested that a continual lipid flow from random sites of insertion into coated pits may carry occupied receptors into coated pits. This model doesn't explain why the receptors of the LDL internalization-defective cells do not get into coated pits. A more likely possibility is that some type of capture step is required. This step may require the activity of a class of enzymes called "transglutaminases" [70, 184].

Using fluorescent derivatives of several different receptor-specific ligands in combination with video intensification microscopy, Willingham, Pastan, and co-workers have observed that the receptors for these ligands are diffusely distributed on the cell surface [reviewed in 70, 184]. The receptors become clustered after ligand binding and are internalized. These observations have been confirmed by electron microscopy for α_2-macroglobulin in mouse fibroblasts [184–187] and epidermal growth factor in human carcinoma A-431 cells [188, 189]. In the case of α_2-macroglobulin, the ligand-receptor complex is partially clustered at 4°C in coated pits, but not after fixation with 0.2% formaldehyde. The ligand-induced clustering appears to be blocked by agents that interfere with transglutaminase activity (see below).

3. Transglutaminase activity and receptor clustering. Enzymes that catalyze an acyl-transfer reaction, where the acyl donor is a peptide-bound

---

Fig. 6. Surface distribution (A, C) and internal location (B, D) of anti-LDL receptor antibodies (A, B) and anti β_2-microglobulin (C, D) after 10 min (A, B) and 15 min (C, D) of exposure to the respective antibodies at 37°C. Normal human fibroblasts were grown on coverslips as previously described [104]. Following the induction of LDL receptors, each set of cells was incubated with 0.2 mg/ml of rabbit anti-LDL receptor [31] (A, B) or 0.2 mg/ml of rabbit anti β_2-microglobulin (BioRad Laboratories) (C, D) for 10 min and 15 min, respectively, at 37°C. One set of cells (B, D) was permeabilized with 0.05% Triton X-100 at -10°C and then treated with 0.5 mg/ml of FITC conjugated antirabbit IgG (Melloy) for 1 h at 37°C. The second set (A, C) was either treated with 0.2 mg/ml anti-LDL receptor (A) followed by exposure to FITC-antirabbit IgG as above or was treated with FITC-antirabbit IgG alone (C). Within 10 min, the anti-LDL receptor had been internalized (B) and at the same time cleared the receptors from the cell surface [42]. In contrast, very little anti β_2-macroglobulin entered the cell in the 15-min time period (D). Instead, it remained on the cell surface (C) (A, B reprinted with permission from Anderson et al: J Cell Biol 93:523–531, 1982.) $\times$ 1,000.

glutamine, are termed "transglutaminases" [190]. Usually the ε amino group of the peptide-bound lysine residue serves as an acceptor to form a covalent ε-(γ-glutamyl) lysine cross link. Transglutaminases are widely distributed in cells, but their biological function is poorly understood. In a recent review by Folk [190], he pointed out that "although there is little doubt that each of the transglutaminases is capable of catalyzing ε-(γ-glutamyl) lysine cross-link formation, available data did not necessarily prove this to be the biological role of each . . . ."

There are two physiological events that clearly depend on transglutaminase-catalyzed reactions: the cross-linking of the fibrin clot during hemostasis, and the production of the vaginal plug by rodent seminal plasma. In both of these cases, the presence of the ε-(γ-glutamyl) lysine linkage has been established and the enzyme appears to be involved in cross-linking protein monomers to form a polymerized network. It is only the ability to identify the presence of this linkage in these two "clotting" reactions that establishes this as an important biological activity for this enzyme.

In 1979, Maxfield et al [191] reported that a variety of amines inhibit the clustering of both α_2-macroglobulin and epidermal growth factor (EGF) on the surfaces of fibroblasts. In addition, they found that the presence of low calcium in the medium prevented clustering. These studies were interpreted to mean that because amines act as competitive substrates or inhibitors of transglutaminase, the activity of this calcium-dependent enzyme was required for ligand-induced clustering. Other transglutaminase inhibitors were also found to prevent clustering and internalization [192]. Furthermore, a correlation existed between the inhibitory activity of these agents and their ability to prevent clustering. Levitzki et al [193] reported that transglutaminase inhibitors other than amines, such as N-benzyloxycarbonyl-5-diazo-4-oxo-norvaline P-nitrophenyl ester (BDONV), were also able to inhibit clustering and internalization of α_2-macroglobulin. These studies primarily used image-intensifying light microscopy to measure the clustering of fluorescent ligand.

Fluorescent-microscopic techniques lack the quantitative precision that is available using radioactive ligands. Therefore, biochemical studies were carried out to test the effects of these inhibitors on ^{125}I-EGF [194, 195] and ^{125}I-α_2-macroglobulin [196] binding and endocytosis. In these studies, the transglutaminase inhibitors bacitracin, methylamine, and diazooxonorvaline did not affect the rate at which ^{125}I-EGF was internalized [194] although dansylcadaverine was inhibitory [195]. The results of these studies were not in agreement with those of Davies et al [192] or Maxfield et al [191] from this laboratory. Haigler et al [194] speculated that the discrepancy in their data had to do with a contamination in the preparation of commercial EGF used to make the rhodamine-lactalbumin-EGF conjugate for the previous studies. Dickson et al [196] studied the effects of these inhibitors on binding and uptake of ^{125}I-α_2-macroglobulin in cultured fibroblasts. Both low calcium and bacitracin inhibited binding of α_2-macroglobulin at low concentrations (0.5 μg/ml) of α_2-macroglobulin, but not at high concentrations (100 μg/ml). Furthermore, following a preincubation at 37°C with inhibitors, low concentrations (0.5 μg/ml) of α_2-macroglobulin were unable to bind in the presence of bacitracin, monodansylcadaverine, and BDONV. These inhibitors had no effect on binding at high concentrations (100 μg/ml) of α_2-macroglobulin. These investigators concluded that these inhibitors had a selective inhibitory effect on high-affinity receptors (detected at 0.5 μg/ml) for α_2-macroglobulin. They proposed that the receptors in coated pits were the high-affinity binding sites and that the low-affinity binding sites were diffusely distributed on the cell surface. The low-affinity sites normally move into coated pits after ligand binding, but transglutaminase inhibitors prevent this step. Although they proposed that low-affinity sites could become high-affinity sites once in the coated pit, no direct evidence for this interconversion was presented.

Other investigators have found that transglutaminase inhibitors appear to affect ligand-receptor internalization. The internalization of Pseudomonas toxin has been reported to be inhibited by methylamine [197]. Agonist-mediated internalization of β-adrenergic receptors in frog erythrocytes is also prevented by this amine [198]. Vitellogenin uptake by Xenopus oocytes may involve transglutaminases [199]. Finally, in fibroblasts, dansylcadaverine appears to prevent insulin-induced receptor loss [200].

Willingham and Pastan have amassed a considerable amount of data to support the idea that cellular transglutaminases play a role in receptor-mediated endocytosis. According to their model, after ligand binding the receptor is susceptible to cross linking by some factor located in coated pits by a transglutaminase-dependent reaction. Receptors move about randomly until they reach a coated pit, where they are captured as a result of the covalent attachment of the receptor to the coated pit. The capture step eventually leads to receptor clustering, which is required for internalization. Although we agree that a capture step may be necessary, other data suggest that transglutaminases are not involved.

Contrary to the results using fluorescent EGF, ^{125}I-EGF internalization is not inhibited by any of the transglutaminase inhibitors except dansylcadaverine [195]. Likewise, McKanna et al [189] found that in A-431 cells, EGF-ferritin conjugates were internalized in the presence of 30 mM methylamine. Yarden et al [201] and King et al [202] also showed that EGF and its receptor are internalized in the presence of amines. These studies suggest that transglutaminases are not involved in the clustering step prior to the internalization of this hormone. Dansylcadavarine could easily be affecting some other aspect of the internalization process.

Interpretation of the effects of these inhibitors on α_2-macroglobulin internalization is made difficult by the fact that other investigators report that α_2-macroglobulin binds to fibroblasts [203, 204] and macrophages [44, 205, 206] only when it is complexed to proteases. Therefore, even though Pastan and Willingham [70] report that there is no difference in the binding of the complexed vs uncomplexed form of α_2-macroglobulin, it is not clear whether these two ligands bind to the same receptor. Furthermore, the relationship between high-affinity and low-affinity receptors for α_2-macroglobulin is poorly understood.

Internalization of radiolabeled α_2-macroglobulin protease complex is not affected by transglutaminase inhibitors in either fibroblasts [207] or macrophages [208]. Rather, these agents appear to inhibit the recycling of this receptor. In fact, careful analysis of the studies by Dickson et al [196] suggests that these agents either prevent the recycling of the high-affinity receptor for α_2-macroglobulin or prevent ligand binding to this receptor. Therefore, it can not be unequivocally concluded that transglutaminase activity is required for the normal function of this class of receptors.

In conclusion, a variety of transglutaminase inhibitors affect the function of several different receptors. However, these agents may not affect all of the receptors in the same way. More importantly, receptor-coated pit cross linking by transglutaminases has not been definitively established. Investigators must demonstrate that this enzyme resides in coated pits and that ϵ-(γ-glutamyl) lysine isopeptide bonds are formed between receptors and some element of the coated pit during internalization. If these criteria can be established, then future studies must determine why multiple receptors that can internalize through the same endocytic vesicle [209] are differentially effected by transglutaminase inhibitors.

In our opinion, the molecular mechanism involved in receptor-coated-pit interactions has yet to be determined. Future research should be directed toward establishing an in vitro model for studying the interaction of these molecular components. Alternatively, the development of a method for removing coated pits from the cell surface to see what happens to receptor distribution and ligand internalization would be useful.

IV. SORTING, RECEPTOR RECYCLING, TARGETING

With few exceptions [210], receptor-mediated endocytosis takes place through coated pits. Despite this ubiquitous uptake system, the final destination of the endocytic vesicles formed from coated pits varies with the ligand and the cell. By and large, most of the receptor-mediated processes studied deliver the ligand to the lysosomal compartment where it is subsequently degraded [reviewed in 7, 211]. In some situations, the ligand is delivered to storage granules [78] or to an extracellular site. The latter process can involve ligand transport from the apical region of the cell to either the lateral border of the cell [127, 212] or the basal border of the cell [64, 213]. The intracellular movement of ligands to selective regions of the cell depends on factors that regulate the interaction of the transport vesicle with specific membrane compartments and upon the sorting of vesicle contents. At the same time, receptors and their ligands often separate so that the former returns to the cell surface (recycling) and the latter goes to a specific intracellular target. The mechanisms involved in achieving targeting, sorting, and receptor recycling are poorly understood, and therefore represent an important area for future research.

An understanding of these processes must begin with an appreciation of the morphological events that accompany ligand uptake. The general features of ligand internalization can be studied by exposing tissue culture cells to a fluorescent ligand. For example, the time sequence of fluorescent LDL internalization is seen in Figure 7. Initially, LDL is bound in linear clusters on the cell surface (Fig. 7A). After 3 min, numerous, larger clusters are seen in regions that subtend the initial sites of LDL binding. Since these vesicles appear larger and less numerous than the initial, surface-bound clusters, most likely some of the newly formed endocytic vesicles have fused at this time point. With further incubation at 37°C (Fig. 7C), the vesicles increase in size and migrate to the perinuclear region of the cell. These pictures show clearly that multiple vesicle fusions take place as the LDL moves into the lysosomal compartment.

Many investigators have described the ultrastructural sequence of receptor-mediated endocytosis using ligands that are labeled with electron-dense probes. Typically they find that the coated pit transforms into a coated vesicle, which rapidly loses its coat material. The rapid loss of coat material may indicate that these are unstable structures. In fact, coated endocytotic vesicles seem to be difficult to isolate by the usual methods [91]. Following the initial vesicle formation, the vesicles get larger and more irregularly shaped. These secondary vesicles seem to have more ligand, which agrees with the light-microscopic data. Eventually these vesicles move to their intracellular target, which in most cases is the lysosome. There is some disagreement about the route these vesicles take and the time required to reach the lysosome. LDL-ferritin conjugates appear in lysosome structures as early as 5–10 min after internalization [125, 211]. α_2-Macroglobulin-HRP conjugates have been reported to take much longer to reach the lysosome and to appear first in the Golgi region of the cell before the lysosome [185, 187].

This morphological information shows that endocytic vesicles undergo multiple fusions before ultimately reaching their intracellular target. These

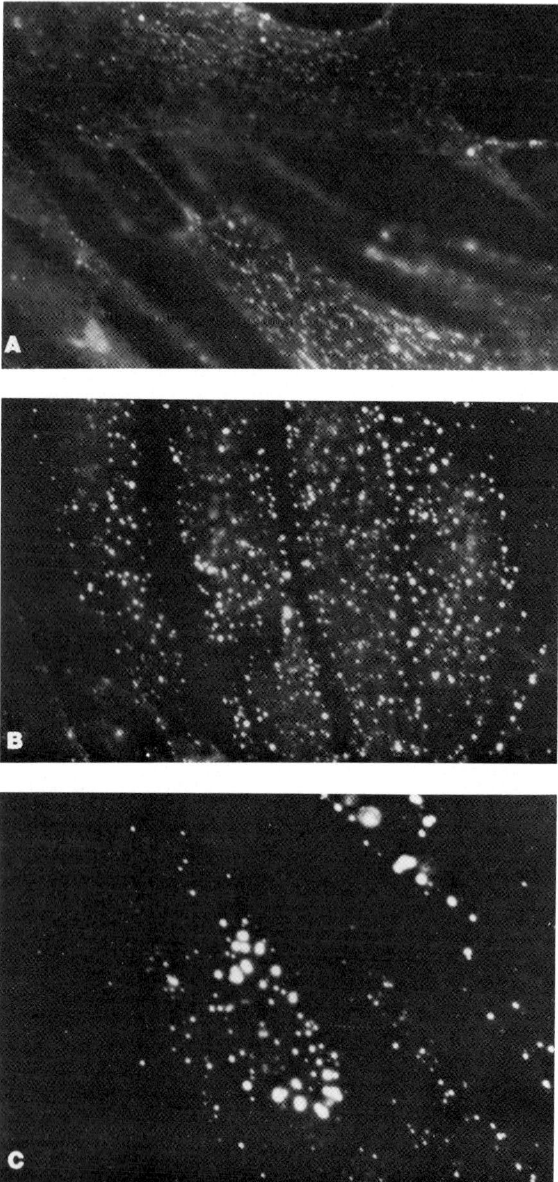

Fig. 7. Visualization by fluorescence microscopy of the internalization of fluorescent LDL. A) Fluorescent LDL on the surfaces of cells after incubation at 4°C for 2 h. B) Cells that had been exposed to fluorescent LDL at 4°C, washed, and warmed for 3 min to 37°C. C) Cells that had been exposed to fluorescent LDL at 4°C, washed, and then warmed to 37°C for 20 min. Note the progressive accumulation of fluorescent LDL into larger and larger vacuoles with time at 37°C. (Reprinted with permission from Anderson et al: J Receptor Res 1:17–39, 1980.) × 1,000.

vesicle-vesicle interactions appear to be specific because under certain conditions they can be inhibited. Dunn et al [156] showed that below 20°C, endocytosis of asialofetuin takes place at a slower rate; however, even though the ligand enters the cell, it never gets degraded—the normal fate of this protein when it reaches the lysosome. This has been interpreted to mean that at this temperature endocytic vesicles do not fuse with the lysosome even though multiple fusions between endocytic vesicles takes place (^{125}I-asialofetuin accumulates in vesicles that are larger than the initial endocytic vesicle [see Fig. 4B, Ref. 156]). These temperature studies have permitted the separation of two different vesicle fusion events that undoubtedly depend on the differential properties of the membranes in these compartments. Of interest is that the presence of certain substances in the endocytic vesicle—for example, concanavalin A [214], ribonuclease A [215], and some types of parasites [216, 217]—inhibits fusion with the lysosome. These agents may act to reorganize transmembrane proteins involved in fusion so that they are no longer able to interact with the lysosome.

These various membrane-membrane interactions represent sites where membrane can be partitioned into separated compartments and at the same time the contents of the vesicle can be segregated for delivery to different intracellular locations. Both membrane partitioning and ligand sorting are necessary events if the receptor is to recycle and if the vesicle contents are to reach their proper target.

A. Sorting

There is now good evidence that two or more macromolecules can enter cells through a common endocytic vesicle [54, 186, 209, 218]. Sometimes this involves cointernalization through the same coated region of membrane [54, 186]. Furthermore, two macromolecules that are initially in the same endocytic vesicle may eventually reach quite separate intracellular compartments [54, 218]. To achieve this important physiologic event, the vesicle contents must be segregated into two separate vesicles that have different intracellular destinations.

The eventual fate of an endocytosed macromolecule may depend in part on its ability to bind to the cell surface membrane. Farquhar and co-workers have studied the differential uptake of cationized ferritin (CF), anionic ferritin (AF), and horseradish peroxidase (HRP) in anterior pituitary cells [219] and immunoglobulin-secreting cells [218]. Because the surfaces of most cells are negatively charged, CF binds to the cell surface and is internalized by adsorptive endocytosis. AF and HRP are taken up by bulk-phase endocytosis. In both types of cells, these investigators found that CF appeared in both Golgi and lysosomal compartments shortly after internalization. When bulk-phase endocytosis was followed, the markers went only to the lysosomal compartment. Furthermore, when cells were exposed to both CF and HRP, both markers were endocytosed by the same vesicle, but CF went to the lysosome and the Golgi, whereas the HRP went only to the lysosome. Beside establishing a likely pathway for membrane recycling (in plasma cells or myeloma cells, CF eventually appeared in exocytic vesicles that contained immunoglobulin), these studies also demonstrate that the contents of the vesicle is sorted and delivered to separate compartments. The membrane-bound ligands apparently have intracellular fates different from those of the unbound ligands.

CF undoubtedly binds to many different glycoproteins and glycolipids. However, similar results were obtained by Abrahamson and Rodewald [54] using immunoglobulin G as a marker for a specific membrane receptor and HRP as a bulk-phase marker in neonatal rat intestinal cells. They found that whereas the IgG was transported to the lateral borders of the cell and excreted, the HRP went to the lysosome. Again, some type of sorting event was required to achieve this differential distribution of the two ligands, which were endocytosed initially into the same vesicle. These investigators postulate that the sorting event takes place before the lysosome.

These studies suggest that membrane-bound ligands tend to remain with the vesicle membrane as it moves to various cellular compartments. One must assume that at some stage, either the primary endocytic vesicle divides or some portion of its membrane buds off to form a new vesicle (illustrated in Fig. 8). If this new vesicle is small compared to the initial vesicle (Fig. 8, I), then it will not contain much content marker, but it will contain a significant amount of membrane-bound marker. As part of this step, the bound marker may cluster to one side of the primary vesicle. The result would be two vesicles, one containing bulk markers (Fig. 8, II), the other bound markers (Fig. 8, I). Each of these vesicles would then be able to go to different parts of the cell.

If this is the basic event that leads to sorting, then the direction a particular bound ligand takes may depend on the ionic conditions of the primary endocytic vesicle (see Fig. 8). Consider the uptake of LDL versus the uptake of IgG in the two types of cells that contain receptors for these ligands. Both are internalized by coated vesicles. Whereas LDL goes to the lysosome, IgG undergoes transcellular transport. At a low pH, LDL dissociates from its receptor [29] but IgG is more tightly bound to its receptor [51]. Thus, if the initial endocytic compartment has a low pH in both cell types [220], then LDL would dissociate from its receptor and act like a bulk-phase marker but IgG would remain bound. The LDL receptor would then be free to return to the cell surface, but the IgG would remain with its receptor as it goes to the lateral cell surface. Cationized ferritin would behave like the IgG; so would antibodies directed against cell-surface determinants (eg, antiamino-peptidase [212] and antiplasma membrane [221–223] antibodies) and certain membrane proteins [224].

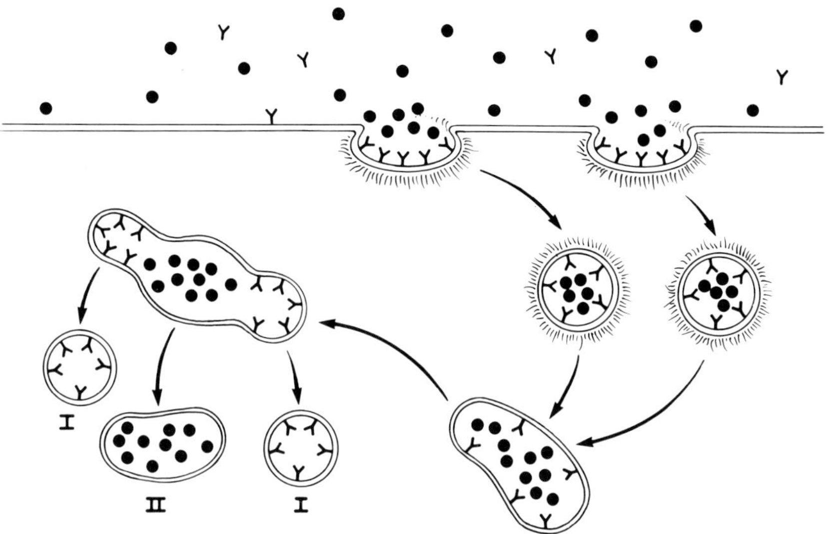

Fig. 8. Diagrammatic representation of the mechanisms involved in the segregation of membrane-bound molecules (Y) from non-membrane-bound molecules (●) following endocytosis. Both membrane-bound and non-membrane-bound molecules can enter through the same coated pit. Once cointernalization into the same coated vesicle takes place, the coat is lost and multiple vesicles fuse to form an intermediate endocytic vesicle. Within this vesicle, membrane-bound molecules segregate and bud off to form one set of vesicles (I). The nonbound molecules remain to form a second set of vesicles (II). These two types of vesicles then move to different regions of the cell.

Although this is a convenient model for sorting, there is evidence that the mechanisms probably differ from cell to cell. For example, Herzog and Farquhar [225] found that dextran, a bulk-phase marker, goes to both lysosomes and Golgi compartments following endocytosis in lacrimal and parotid gland cells. Goud et al [226] showed that membrane-bound and nonbound markers both went to the Golgi. Herzog and Reggio [227] established that two bulk-phase markers, dextran and HRP, were transported to different compartments—dextran to the Golgi and HRP to the lysosome. Van Duers et al [228] found that in choroid plexus epithelial cells, native ferritin goes to the lysosome but cationized ferritin goes primarily to the basolateral cell membrane. Finally, in adenohyphyseal cells, HRP is transported to the Golgi compartment [229].

B. Receptor Recycling

To direct two different molecules taken up by the same endocytic vesicle to different intracellular compartments, the ligands must become segregated.

Similarly, if a membrane receptor is to recycle during endocytosis, at some point the ligand and the receptor must become separated so that the latter can return to the cell surface and the former go to its intracellular destination. Both sorting and recycling may involve similar sites of membrane-membrane interaction.

There are several reasons why investigators have come to the conclusion that some membrane receptors are able to recycle. As discussed before, it is a property of the class II receptors that they are able to internalize continuously over long periods of time, even in the presence of protein synthesis inhibitors [230]. In most cases, the internalized ligand is delivered to the lysosome where it is degraded. Thus, for a cell to continue to internalize molecules, either the receptor must escape degradation and return to the cell surface or there must be an intracellular pool of receptors that can be utilized to replace those lost during internalization and degradation. With the exception of the asialoglycoprotein receptor, most receptors involved in internalizing extracellular molecules do not seem to be abundant within the cell. For example, fewer than 10% of the total receptors for LDL in the human fibroblast are located within an intracellular compartment [42, 231]. Since all of these receptors can continually internalize in the absence of protein synthesis for up to 20 hours, the receptor must be recycled. Most likely the process of receptor recycling is part of a general mechanism for recycling membrane [11, 12].

Even the asialoglycoprotein receptor probably recycles. Ashwell and collaborators noted that more than 95% of the asialoglycoprotein binding sites were inside the cell [232]. Furthermore, they found that many of these binding sites appeared to be exposed to the cytoplasmic surface of the lysosome [232]. Since this is the opposite orientation to what would be expected for a receptor coming from the cell surface, they proposed that after internalization the receptor somehow changes its orientation in the membrane, which causes it to escape lysosomal degradation. Blumenthal et al [233] suggested a mechanism for how this might take place. They were able to reconstitute the asialoglycoprotein receptor into a synthetic membrane monolayer and to show that after ligand binding the receptor changed from an asymmetric orientation in the membrane to a symmetric orientation in the presence of an electrochemical gradient. According to their model, within the cell an electrochemical gradient causes the occupied receptor to flip from one side of the membrane to the other. However, other evidence suggests that the large internal pool of receptors is independent of the surface receptors. Stockert et al [234] were able to show that after chemical modification of the asialoglycoprotein receptor there was no exchange of the modified receptor with the internal pool of unmodified receptors. Furthermore, antibodies against this receptor bind and are internalized, but at the same time there is a loss of receptor activity from the cell surface [235]. A similar phenomenon occurs when fibroblasts internalize polyclonal antibodies against the LDL receptor [42]; in both cases, once the surface receptors are cleared from the cell surface they are not replaced by the recruitment of an intracellular pool. Doyle et al [236] have been able to reconstitute the isolated receptor into liposomes and transfer it to cells that are normally devoid of these receptors. Even though these cells do not have an internal pool of receptors, they are able to continuously take up and degrade asialoglycoproteins for an extended period of time. Finally, Geuze et al [49] recently showed by immunocyto-chemical labeling of thin sections that this receptor appears to recycle from a compartment that is near the cell surface.

Based on similar kinds of experiments, many receptors have been found to recycle [207, 208, 230, 236–243]. The fundamental question emerging from all these studies is how these receptors escape degradation and return to the cell surface. Two possible mechanisms come to mind: Either the receptor somehow has special properties that prevent it from being degraded in the lysosome after reaching this compartment, or the receptor never reaches the lysosome.

Observations from the study of other receptor systems suggest that in some cases membrane proteins can be transported through the lysosomal com-partment without being hydrolyzed. Schneider et al [221, 222] demonstrated that monovalent antibodies directed against plasma membrane antigens can bind to the membrane, move to the lysosomal compartment and return to the cell surface without being degraded. Thyberg et al [244] obtained similar results using cationized ferritin as a probe. Another observation from the study of macrophages fits this idea. Polystyrene beads coated with lactoper-oxidase are phagocytosed by macrophages and delivered to a lysosome-enriched membrane compartment, the phagolysosome. Upon the addition of hydrogen peroxide and ^{125}I to these cells at 0°C, the membranes surrounding the phagocytic compartment become radioiodinated. After shifting from 0°C to 37°C, the iodinated membrane returns to the cell surface [245, 246]. Various control experiments indicate that the appearance of the cell surface radioactivity was due to direct recycling of the internalized membrane. Even though these studies show that plasma membrane components can exist intact in a lysosomal environment, none of these studies has determined the effi-ciency of recycling. Conceivably a sizable fraction of either membrane pro-teins or membrane-bound proteins do get degraded [see, eg, 247].

Even though there is a precedence for the idea that membrane determinants or membrane-bound ligands can go to the lysosome without being degraded, no one has determined whether any of the receptors involved in receptor-mediated endocytosis go to this compartment. There is evidence, however, that many receptors do not recycle in the presence of agents such as chloroquine [240, 241, 243], monensin [231], and various amines [207, 208], all of which raise the pH of the lysosome. Maybe low pH is important for proper receptor recycling.

From this and other analyses, there apparently are three possible sites for receptor recycling [42, 230]: 1) Receptors may not ever enter the cell; 2) receptors may enter the cell and recycle from an endocytic compartment; or 3) receptors may traffic all the way to the lysosome and then return to the cell surface. Evidence from the studies of Basu et al [231] and Weigel [238] suggests that receptors do enter the cell during recycling. However, no one has yet determined if receptors recycle from the lysosome or from the endocytic compartment. Since agents that raise the lysosomal pH interrupt recycling, possibly all of these receptors actually traffic through the lysosome. However, recent evidence by Maxfield and Tycko [220] suggests that the initial endocytic vesicle has a low pH. Also, the pH of the initial phagocytic compartment transiently rises to pH 7.4 and then drops to pH 5.4 within 15–20 min in the presence of agents that prevent phagosome-lysosome fusion [248]. Thus, drugs that raise the pH could also be working at the level of the initial endocytic compartment. Since there is evidence for ligand sorting in a prelysosomal compartment (see discussion above), this could also be the compartment where receptors are separated from ligand during recycling. This makes sense because it often takes 10 minutes or longer for ligands to reach a degradative compartment, even though recycling occurs much faster.

C. Targeting

Much more needs to be learned about sorting and receptor recycling. But even the models outlined above do not explain how the cell is able to target endocytic vesicles or secondary vesicles to specific compartments within the cell. Why, for example, does a vesicle that contains LDL go to a lysosome but one that contains IgG go to the lateral border? Maybe vesicle membranes contain cytoplasmically oriented determinants that recognize a particular subset of membranes. Alternatively, maybe fusion is controlled by cytoplasmic factors that recognize and bind together only those vesicles with the right properties. For example, Pollard and co-workers have shown that a cytoplasmic protein, synexin, is able to cause chromaffin granules to aggregate with each other in a calcium-dependent reaction [249, 250]. Creutz [251] then showed that after chromaffin granule aggregation, they fuse in response to the presence of cis-unsaturated fatty acids such as arachidonic acid. Conceivably there is a whole set of synexin-like molecules that are important for specifying the eventual cytoplasmic compartment a given vesicle will reach. Certainly, further defining these important intracellular targeting events will be an important future area of investigation.

V. CONCLUSION

Receptor-mediated endocytosis is an exciting area of research. From its meager beginning as a series of morphological observations, it is now recognized to be a fundamentally important cellular activity. As with all scientific endeavors, there is much more to be learned. Receptor recycling, mechanisms underlying receptor-coated-pit interaction, ligand sorting and targeting, and the molecular basis of endocytosis are topics that deserve attention in future studies.

ACKNOWLEDGMENTS

We are indebted to Ms. Mary Surovik for her enormous effort during the preparation of this manuscript.

VI. REFERENCES

1. Overton E: Uber die osmotischen eigenschafter der lebenden pflanzenund tierzelle vjseher. Naturforsch Ges Zurich 40:159–201, 1895.
2. Hendler, RW: Biological membrane ultrastructure. Physiol Rev 51:66–97, 1971.
3. Silverstein SC, Steinman RM, Cohn ZA: Endocytosis. Annu Rev Biochem 46:669–722, 1977.
4. Steinman RM, Cohn ZA: The interaction of soluble horseradish peroxidase with mouse peritoneal macrophages in vitro. J Cell Biol 55:186–204, 1972.
5. Steinman RM, Silver JM, Cohn ZA: Pinocytosis in fibroblasts: Quantitative studies in vitro. J Cell Biol 63:949–969, 1974.
6. Steinman RM, Brodie SE, Cohn ZA: Membrane flow during pinocytosis. J Cell Biol 68:665–687, 1976.
7. Goldstein JL, Anderson RGW, Brown MS: Coated pits, coated vesicles, and receptor-mediated endocytosis. Nature 279:679–685, 1979.
8. Goldstein JL, Anderson RGW, Brown MS: Receptor-mediated endocytosis and the cellular uptake of low density lipoprotein. In "Membrane Recycling," Ciba Foundation Symposium 92:77–95, 1982.
9. Neville DM, Chang T-M: Receptor-mediated protein transport into cells. Entry mechanisms for toxins, hormones, antibodies, viruses, lysosomal hydrolases, asialoglycoproteins, and carrier proteins. In Bonner F, Kleinzeller A (eds): "Current Topics in Membranes and Transport." New York: Academic Press, 1976, Vol 10, pp 65–150.
10. King AC, Cuatrecasas P: Peptide hormone-induced receptor mobility, aggregation, and internalization. N Engl J Med 305:77–88, 1981.
11. Holtzman E, Mercurio AM: Membrane circulation in neurons and photoreceptors: Some unresolved issues. In Bourne GH, Danielli JF (eds): "International Review of Cytology." New York: Academic Press, Vol 67, 1980, pp 1–67.
12. Waksman A, Hubert P, Cremel G, Rendon A, Burgun C: Translocation of proteins through biological membranes: A critical view. Biochim Biophys Acta 604:249–296, 1980.
13. Hudgin RL, Pricer WE, Ashwell G, Stockert RT, Morell AG: Isolation and properties of a rabbit liver binding protein specific for asialoglycoproteins. J Biol Chem 249:5536–5543, 1974.

14. Innerarity TL, Kempner ES, Hui DY, Mahley RW: Functional unit of the low density lipoprotein receptor of fibroblasts: A 100,000 dalton structure with multiple binding sites. Proc Natl Acad Sci USA 78:4379–4382, 1981.

15. Steer CJ, Kempner ES, Ashwell G: Molecular size of the hepatic receptor for asialoglycoproteins determined in situ by radiation inactivation. J Biol Chem 256:5851–5856, 1981.

16. Schneider WJ, Beisiegel U, Goldstein JL, Brown MS: Purification of the low density lipoprotein receptor: An acidic glycoprotein of 164,000 molecular weight. J Biol Chem 257:2664–2673, 1982.

17. Drickamer K: Complete amino acid sequence of a membrane receptor for glycoproteins: Sequence of the chicken hepatic lectin. J Biol Chem 256:5827–5839, 1981.

18. Saviolakis GA, Harrison LC, Roth J: The binding of ^{125}I-insulin to specific receptors in IM-9 human lymphocytes: Detection of radioactivity covalently linked to receptors. J Biol Chem 256:4924–4928, 1981.

19. Hock RA, Nexo E, Hollenberg MD: Isolation of the human placenta receptor for epidermal growth factor urogastrone. Nature 277:403–405, 1979.

20. Baker JB, Simmer RL, Glenn KC, Cunningham DD: Thrombin and epidermal growth factor become linked to cell surface receptors during mitogenic stimulation. Nature 278:743–745, 1979.

21. Linsley PS, Blifeld C, Wrann M, Fox CF: Direct linkage of epidermal growth factor to its receptor. Nature 278:745–748, 1979.

22. Issacs JD, Savion N, Gospodarowicz D, Fenton JW, Shuman MA: Covalent binding of thrombin to specific sites on corneal endothelial cells. Biochemistry 20:398–403, 1981.

23. Comens PG, Simmer RL, Baker JB: Direct linkage of ^{125}I-EGF to cell surfaces: A useful artifact of chloramine T treatment. J Biol Chem 257:42–45, 1982.

24. Carpenter G, Cohen S: ^{125}I-labelled human epidermal growth factor: Binding, internalization and degradation in human fibroblast. J Cell Biol 71:159–171, 1976.

25. Ascoli M, Puett D: Degradation of receptor bound human choriogonadotropin by murine Leydig tumor cells. J Biol Chem 253:4892–4899, 1978.

26. Wiley HS, Cunningham DD: A steady state model for analyzing the cellular binding, internalization and degradation of polypeptide ligands. Cell 25:433–440, 1981.

27. Kaplan J: Polypeptide-binding membrane receptors: Analysis and classification. Science 212:14–20, 1981.

28. Goldstein JL, Brown MS: Familial hypercholesterolemia: Identification of a defect in the regulation of 3-hydroxy-3-methylglutaryl coenzyme A reductase activity associated with overproduction of cholesterol. Proc Natl Acad Sci USA 70:2804–2808, 1973.

29. Goldstein JL, Brown MS: The low-density lipoprotein pathway and its relation to atherosclerosis. Ann Rev Biochem 46:897–930, 1977.

30. Ashwell G, Morell AG: The role of surface carbohydrates in the hepatic recognition and transport of circulating glycoproteins. Adv Enzymol 41:99–128, 1974.

31. Neufeld EF, Lim TW, Shapiro LJ: Inherited disorders of lysosomal metabolism. Ann Rev Biochem 44:357–376, 1975.

32. Levitzki A: Quantitative aspects of ligand binding to receptors. In Schulster D, Levitzki A (eds): "Cellular Receptors for Hormones and NeuroTransmitters." New York: Wiley, 1980, pp 9–28.

33. Pollet FJ, Levey GS: Principles of membrane receptor physiology and their application to clinical medicine. Ann Intern Med 92:663–680, 1980.

34. Gardner JD: Receptors for gastrointestinal hormones. Gastroenterology 76:202–214, 1979.

35. Segaloff DL, Ascoli M: Removal of the surface-bound human choriogonadotropin results in the cessation of hormonal responses in cultured Leydig tumor cells. J Biol Chem 256:11420–11423, 1981.

36. Hizuka N, Gorden P, Lesniak MA, Van Obberghen E, Carpentier JL, Orci L: Polypeptide hormone degradation and receptor regulation are coupled to ligand internalization. J Biol Chem 256:4591–4597, 1980.

37. Lesniak MA, Roth J: Regulation of receptor concentration by homologous hormone. Effect of human growth hormone or its receptor in IM-9 lymphocytes. J Biol Chem 251:3720–3729, 1976.

38. Gammeltoft S, Kritensen LO, Sestoft L: Insulin receptors in isolated rat hepatocytes: Reassessment of binding properties and observations on the inactivation of insulin at 37°C. J Biol Chem 253:8406–8413, 1978.

39. Sullivan SJ, Zigmond SH: Chemotactic receptor modulation in polymorphonuclear leukocytes. J Cell Biol 85:703–711, 1980.

40. Zigmond SH, Sullivan SJ, Lauffenburger DA: Kinetic analysis of chemotactic peptide modulation. J Cell Biol 92:34–43, 1982.

41. Anderson RGW, Goldstein JL, Brown MS: Localization of low density lipoprotein receptors on plasma membrane of normal human fibroblasts and their absence in cells from a familial hypercholesterolemia homozygote. Proc Natl Acad Sci USA 73:2434–2438, 1976.

42. Anderson RGW, Brown MS, Beisiegel U, Goldstein JL: Surface distribution and recycling of the LDL receptor as visualized with anti-receptor antibodies. J Cell Biol 93:523–531, 1982.

43. Wall DA, Hubbard AL: Galactose-specific recognition system of mammalian liver: Receptor distribution on the hepatocyte cell surface. J Cell Biol 90:687–696, 1981.

44. Kaplan J, Kaplan S: The native distribution of receptors for α-macroglobulin-protease complexes and mannose terminal glycoproteins on the surface of alveolar macrophages is non-random. In preparation, 1982.

45. Roth TF, Cutting JA, Atlas SB: Protein transport: A selective membrane mechanism. J Supramol Struc 4:527–548, 1976.

46. Willingham MC, Pastan IH, Sahagian GG, Jourdian GW, Neufeld EF: Morphologic study of the internalization of a lysosomal enzyme by the mannose-6-phosphate receptor in cultured chinese hamster ovary cells. Proc Natl Acad Sci USA 78:6967–6971, 1981.

47. Fawcett DW: Surface specializations of absorbing cells. J Histochem Cytochem 13:75–91, 1965.

48. Stockert RJ, Haimes HB, Morell AG, Novikoff PM, Novikoff AB, Quintana N, Sternlieb I: Endocytosis of asialoglycoprotein-enzyme conjugates by hepatocytes. Lab Invest 43:556–563, 1980.

49. Geuze HJ, Slot JW, Strous GJAM, Lodish HF, Schwartz AL: Immunocytochemical localization of the receptor for asialoglycoprotein in rat liver cells. J Cell Biol 92:865–870, 1982.

50. Rome LH, Weissmann B, Neufeld EF: Direct demonstration of binding of a lysosomal enzyme, α-L-iduronidase, to receptors on cultured fibroblasts. Proc Natl Acad Sci USA 76:2331–2334, 1979.

51. Rodewald R: pH-dependent binding of immunoglobulins to the intestinal cells of the neonatal rat. J Cell Biol 71:666–670, 1976.

52. Rodewald R: Immunoglobulin transmission in mammalian young and the involvement of coated vesicles. In Ockleford CD, Whyte A (eds): "Coated Vesicles." Cambridge: Cambridge University Press, 1980, pp 69–101.

53. Borthistle BK, Kubo RT, Brown WR, Grey HM: Studies on receptors for IgG on epithelial cells of the rat intestine. J Immunol 119:471–476, 1977.

54. Abrahamson DR, Rodewald R: Evidence for the sorting of endocytic vesicle contents during the receptor-mediated transport of IgG across the newborn rat intestine. J Cell Biol 91:270–280, 1981.

55. Rodewald R: Distribution of immunoglobulin G receptors in the small intestine of the young rat. J Cell Biol 85:18–32, 1980.
56. Morgan EH: Comparative iron metabolism. In Jacobs A, Wordwood M (eds): "Biochemistry and Medicine." London: Academic Press, 1974, pp 641–687.
57. Glass J, Nunez MT, Robinson SH: Iron transport from sepharose-bound transferrin. Biochem Biophys Res Commun 75:226–232, 1977.
58. Brown MS, Goldstein JL: Regulation of the activity of the low density lipoprotein receptor in human fibroblasts. Cell 6:307–316, 1975.
59. Ward JH, Kushner JPK, Kaplan J: Regulation of HeLa cell transferrin receptor. J Biol Chem 257:10317–10323, 1982.
60. Gardner JM, Fambrough DM: Acetylcholine receptor degradation measured by density labelling: Effects of cholinergic ligands and evidence against recycling. Cell 16:661–674, 1979.
61. Landau EM: Function and structure of the Ach receptor at the muscle endplate. Prog Neurobiol 10:253–288, 1978.
62. Mukherjee C, Caron MG, Lefkowitz RJ: Catecholamine-induced subsensitivity of adenylate cyclase associated with loss of β-adrenergic receptor binding sites. Proc Natl Acad Sci USA 72:1945–1949, 1975.
63. Lad PM, Nielsen TB, Preston MS, Rodbell M: The role of the guanine nucleotide exchange reaction in the regulation of the β-adrenergic receptor and in the actions of catecholamines and the cholera toxin on adenylate cyclase in turkey erythrocyte membranes. J Biol Chem 255:988–995, 1980.
64. Anderson RGW: Cell surface membrane structure and the function of endothelial cells. In Schwartz CJ, Werthessen NT, Wolf S (eds): "Structure and Function of the Circulation." New York: Plenum Press, 1981, pp 239–286.
65. Lewis WH: Pinocytosis. Bull Johns Hopkins Hosp 49:17–36, 1931.
66. Holter H: Pinocytosis. Int Rev Cytol 8:481–504, 1959.
67. Willingham MC, Yamada SS: A mechanism for the destruction of pinosomes in cultured fibroblasts. Piranhalysis. J Cell Biol 78:480–487, 1978.
68. Palade GE: The endoplasmic reticulum. J. Biophys Biochem Cytol 2:85–97, 1956, Suppl 4.
69. Bennett HS: The concepts of membrane flow and membrane vesiculation as mechanisms for active transport and ion pumping. J Biophys Biochem Cytol 2:99–103, 1956, Suppl 4.
70. Pastan IH, Willingham MC: Receptor-mediated endocytosis of hormones in cultured cells. Ann Rev Physiol 43:239–250, 1981.
71. Wagner RC, Casley-Smith JR: Endothelial vesicles. Microvasc Res 21:267–298, 1981.
72. Palade GE: Fine structure of blood capillaries. J Appl Physiol A24:1424, 1953.
73. Palade GE: Blood capillaries of the heart and other organs. Circulation 24:368–384, 1961.
74. Bundgaard M, Frokjaer-Jensen J, Crone C: Endothelial plasmalemmal vesicles as elements in a system of branching invaginations from the cell surface. Proc Natl Acad Sci USA 76:6439–6442, 1979.
75. Mugnaini E, Osen KK, Schnapp B, Friedrich VL Jr: Distribution of Schwann cell cytoplasm and plasmalemmal vesicles (caveolae) in peripheral myelin sheaths: An electron microscopic study with thin sections and freeze-fracturing. J Neurocytol 6:647–668, 1977.
76. Bretscher MS, Whytock S: Membrane-associated vesicles in fibroblasts. J Ultrastruct Res 61:215–217, 1977.
77. Rosenbluth J, Wissig SL: The distribution of exogenous ferritin in toad spinal ganglia and the mechanism of its uptake by neurons. J Cell Biol 23:307–325, 1964.

78. Roth TF, Porter KR: Yolk protein uptake in the oocyte of the mosquito Aedes aegypti L. J Cell Biol 20:313–332, 1964.
79. Policard A, Bessis M: Sur un mode d'incorporation des macromolecules par la cellule, visible au microscope electronique; la rhopheocytose. Comptes Rendus Hebdomadaire des Seances de l'Academie des Sciences 246:3194, 1958.
80. Bessis M, Breton-Gorius J: Differents aspects du fer dans l'organisme. I. Ferritine et micelles ferrugineuses. J Biophys Biochem Cytol 6:231–240, 1959.
81. Gray EG: The granule cells mossy synapses and Purkinje spine synapses of the cerebellum: Light and electron microscopic observations. J Anat 95:345–387, 1961.
82. Palay SL: Alveolate vesicles in Purkinje cells of the rat's cerebellum. J Cell Biol 19:89a, 1963.
83. Maunsbach AB: Electron microscopic observations on ferritin absorption in microperfused renal proximal tubules. J Cell Biol 19:48a, 1963.
84. Maunsbach AB: Absorption of ferritin by rat kidney proximal tubule cells: Electron microscopic observations of the initial uptake phase in cells of microperfused single proximal tubules. J Ultrastruct Res 16:1–12, 1966.
85. Kanaseki T, Kadota K: The "vesicle in a basket": A morphological study of the coated vesicle isolated from the nerve endings of the guinea pig brain, with special reference to the mechanism of membrane movements. J Cell Biol 42:202–220, 1969.
86. Ockleford CD: A three-dimensional reconstruction of the polygonal pattern on placental coated-vesicle membranes. J Cell Sci 21:83–91, 1976.
87. Heuser J, Evans L: Three-dimensional visualization of coated vesicle formation in fibroblasts. J Cell Biol 84:560–583, 1980.
88. Pearse BMF: Coated vesicles from pig brain: Purification and biochemical characterization. J Mol Biol 97:93–98, 1975.
89. Pearse BMF: Clathrin: A unique protein associated with intracellular transfer of membrane by coated vesicles. Proc Natl Acad Sci USA 73:1255–1259, 1976.
90. Pearse BMF: On the structural and functional components of coated vesicles. J Mol Biol 126:803–812, 1978.
91. Pearse B: Coated vesicles. Trends Biochem Sci 5:131–134, 1980.
92. Woods JW, Woodward MP, Roth TF: Common features of coated vesicles from dissimilar tissues: Composition and structure. J Cell Sci 30:87–97, 1978.
93. Crowther RA, Finch JT, Pearse BMF: On the structure of coated vesicles. J Mol Biol 103:785–798, 1976.
94. Blitz AL, Fine RE, Toselli PA: Evidence that coated vesicles isolated from brain are calcium-sequestering organelles resembling sarcoplasmic reticulum. J Cell Biol 75:135–147, 1977.
95. Woodward MP, Roth TF: Influence of buffer ions and divalent cations on coated vesicle disassembly and reassembly. J Supramol Struc 11:237–250, 1979.
96. Keen JH, Willingham MC, Pastan IH: Clathrin-coated vesicles: Isolation, dissociation and factor-dependent reassociation of clathrin baskets. Cell 16:303–312, 1979.
97. Schook W, Puszkin S, Bloom W, Ores C, Kochwa S. Mechanochemical properties of brain clathrin: Interactions with actin and α-actinin and polymerization into basket-like structures or filaments. Proc Natl Acad Sci USA 76:116–120, 1979.
98. Van Jaarsveld PP, Nandi PK, Lippoldt RE, Saroff H, Edelhoch H: Polymerization of clathrin protomers into basket structures. Biochemistry 20:4129–4135, 1981.
99. Nandi PK, Pretorius HT, Lippoldt RE, Johnson ML, Edelhoch H: Molecular properties of the reassembled coat protein of coated vesicles. Biochemistry 19:5917–5921, 1980.
100. Pretorius HT, Nandi PK, Lippoldt RE, Johnson ML, Keen JH, Pastan I, Edelhoch H: Molecular characterization of human clathrin. Biochemistry 20:2777–2782, 1981.
101. Willingham MC, Keen JH, Pastan IH: Ultrastructural immunocytochemical localization of clathrin in cultured fibroblasts. Exp Cell Res 132:329–338, 1981.

102. Cheng TP-O, Byrd FI, Whitaker JN, Wood JG: Immunocytochemical localization of coated vesicle protein in rodent nervous system. J Cell Biol 86:624–633, 1980.
103. Keen JH, Willingham MC, Pastan I: Clathrin and coated vesicle proteins: Immunological characterization. J Biol Chem 256:2538–2544, 1981.
104. Anderson RGW, Vasile E, Mello RJ, Brown MS, Goldstein JL: Immunocytochemical visualization of coated pits and vesicles in human fibroblasts: Relation to low density lipoprotein receptor distribution. Cell 15:919–933, 1978.
105. Fine RE, Blitz AL, Sack DH: Characterization of anti-clathrin serum. FEBS Lett 94:59–62, 1978.
106. Kartenbeck J, Schmid E, Muller H, Franke WW: Immunological identification and localization of clathrin and coated vesicles in cultured cells and in tissues. Exp Cell Res 133:191–211, 1981.
107. Bloom WS, Fields KL, Yen SH, Haver K, Schook W, Puszkin S: Brain clathrin: Immunofluorescent patterns in cultured cells and tissues. Proc Natl Acad Sci USA 77:5520–5524, 1980.
108. Ungewickell E, Branton D: Assembly units of clathrin coats. Nature 289:420–422, 1981.
109. Kirchhausen T, Harrison SC: Protein organization in clathrin trimers. Cell 23:755–761, 1981.
110. Unanue ER, Ungewickell E, Branton D: The binding of clathrin triskelions to membranes from coated vesicles. Cell 26:439–446, 1981.
111. Crowther RA, Pearse BMF: Assembly and packing of clathrin into coats. J Cell Biol 91:790–797, 1981.
112. Korn ED: Cell membranes: Structure and synthesis. Annu Rev Biochem 38:263–288, 1969.
113. Elias PM, Goerke J, Friend DS: Freeze-fracture identification of sterol-digitonin complexes in cell and liposome membranes. J Cell Biol 78:577–596, 1978.
114. Montesano R, Perrelet A, Vassalli P, Orci L: Absence of filipin-sterol complexes from large coated pits on the surface of cultured cells. Proc Natl Acad Sci USA 76:6391–6395, 1979.
115. Robinson JM, Karnovsky MJ: Evaluation of the polyene antibiotic filipin as a cytochemical probe for membrane cholesterol. J Histochem Cytochem 28:161–168, 1980.
116. Orci L, Carpentier J-L, Perrelet A, Anderson RGW, Goldstein JL, Brown MS: Occurrence of low density lipoprotein receptors within large pits on the surface of human fibroblasts as demonstrated by freeze-etching. Exp Cell Res 113:1–13, 1978.
117. Franke WW, Luder MR, Kartenbeck J, Zerban H, Keenan TW: Involvement of vesicle coat material in casein secretion and surface regeneration. J Cell Biol 69:173–195, 1976.
118. Friend DS, Farquhar MG: Functions of coated vesicles during protein absorption in the rat vas deferens. J Cell Biol 35:357–376, 1967.
119. Heuser JE, Reese TS: Evidence for recycling of synaptic vesicle membrane during transmitter release at the frog neuromuscular junction. J Cell Biol 57:315–344, 1973.
120. Maul GG, Brumbaugh JA: On the possible function of coated vesicles in melanogenesis of the regenerating fowl feather. J Cell Biol 48:41–48, 1971.
121. Willingham MC, Pastan I: The receptosome: An intermediate organelle of receptor-mediated endocytosis in cultured fibroblasts. Cell 21:67–77, 1980.
122. Willingham MC, Maxfield FR, Pastan I: Receptor-mediated endocytosis of alpha$_2$-macroglobulin in cultured fibroblasts. J Histochem Cytochem 28:818–823, 1980.
123. Wehland J, Willingham MC, Dickson R, Pastan I: Microinjection of anticlathrin antibodies into fibroblasts does not interfere with the receptor-mediated endocytosis of α_2-macroglobulin. Cell 25:105–119, 1981.

124. Willingham MC, Rutherford AV, Gallo MG, Wehland J, Dickson RB, Schlegel R, Pastan IH: Receptor-mediated endocytosis in cultured fibroblasts: Cryptic coated pits and the formation of receptosomes. J Histochem Cytochem 29:1003–1013, 1981.
125. Anderson RGW, Brown MS, Goldstein JL: Role of the coated endocytic vesicle in the uptake of receptor-bound low density lipoprotein in human fibroblasts. Cell 10:351–364, 1977.
126. Salisbury JL, Condeelis JS, Satir P: Cytoskeletal architecture during receptor-mediated endocytosis. J Cell Biol 91:419a, 1981.
127. Rodewald R: Intestinal transport of antibodies in the newborn rat. J Cell Biol 58:189–211, 1973.
128. Wall DA, Wilson G, Hubbard AL: The galactose-specific recognition system of mammalian liver: The route of ligand internalization in rat hepatocytes. Cell 21:79–93, 1980.
129. Mello RJ, Brown MS, Goldstein JL, Anderson RGW: LDL receptors in coated vesicles isolated from bovine adrenal cortex: Binding sites unmasked by detergent treatment. Cell 20:829–837, 1980.
130. Booth AG, Wilson MJ: Human placental coated vesicles contain receptor-bound transferrin. Biochem J 196:355–362, 1981.
131. Fine RE, Goldenberg R, Sorrentino J, Herschman HR: Subcellular structures involved in internalization and degradation of epidermal growth factor. J Supramol Struc Cell Biochem 15:235–251, 1981.
132. Hewett JA: On the influence of polyvalent ligands on membrane curvature. J Theor Biol 64:455–472, 1977.
133. Singer SJ, Ash JF, Bourguignon LYW, Heggeness MH, Louvard D: Transmembrane interactions and the mechanisms of transport of proteins across membranes. J Supramol Struc 9:373–389, 1978.
134. Deuticke B: Transformation and restoration of biconcave shape of human erythrocytes induced by amphiphilic agents and changes of ionic environment. Biochim Biophys Acta 163:494–500, 1968.
135. Bessis M, Weed RI, Leblond PF (eds): "Red Cell Shape: Physiology, Pathology, Ultrastructure." Berlin: Springer-Verlag, 1973.
136. Sheetz MP, Singer SJ: Biological membranes as bilayer couples: A molecular mechanism of drug-erythrocyte interaction. Proc Natl Acad Sci USA 71:4457–4461, 1974.
137. Sheetz MP, Singer SJ: Equilibrium and kinetic effects of drugs on the shapes of human erythrocytes. J Cell Biol 70:247–251, 1976.
138. Sheetz MP, Singer SJ: On the mechanism of ATP-induced shape changes in human erythrocyte membranes. I. The role of the spectrin complex. J Cell Biol 73:638–646, 1977.
139. Sheetz, MP, Painter RG, Singer SJ: The contractile proteins of erythrocyte membranes and shape changes. In Goldman R, Pollard T, Rosenbaum J (eds): "Cell Motility." Cold Spring Harbor, New York: Cold Spring Harbor Laboratory, 1976, pp 651–664.
140. Sheetz MP, Painter RG, Singer SJ: Biological membranes as bilayer couples. III. Compensatory shape changes induced in membranes. J Cell Biol 70:193–203, 1976.
141. Birchmeier W, Singer SJ: On the mechanism of ATP-induced shape changes in human erythrocyte membranes. II. The role of ATP. J Cell Biol 73:647–659, 1977.
142. Birchmeier W, Singer SJ: Muscle G-actin is an inhibitor of ATP-induced erythrocyte ghost shape changes and endocytosis. Biochem Biophys Res Commun 77:1354–1360, 1977.
143. Evans EA: Bending resistance and chemically induced moments in membrane bilayers. Biophys J 14:923–931, 1974.

144. Brailsford JD, Korpman RA, Bull BS: Crenation and cupping of the red cell: A new theoretical approach. Part II. Cupping. J Theor Biol 86:531–546, 1980.

145. Penniston JT, Green DE: The conformational basis of energy transformations in membrane systems. IV. Energized states and pinocytosis in erythrocyte ghosts. Arch Biochem Biophys 128:339–350, 1968.

146. Hayashi H, Plishker GA, Vaughan L, Penniston JT: Energy-dependent endocytosis in erythrocyte ghosts. IV. Effects of Ca^{++}, Na^{+} + K^{+} and 5′-adenylimidodiphosphate. Biochim Biophys Acta 382:218–229, 1975.

147. Jarrett HW, Reid TB, Penniston JT: Concurrent inhibition of the low-affinity Ca^{++}-stimulated ATPase and Mg^{++} ATP-dependent endocytosis in erythrocyte ghosts by N-napthylmaleimide and carbomylcyanide-m-chlorophenylhydrazone. Arch Biochem Biophys 183:498–510, 1977.

148. Ben-Bassat I, Bensch KG, Schrier SL: Drug-induced erythrocyte membrane internalization. J Clin Invest 51:1833–1844, 1972.

149. Ginn FL, Hochstein P, Trump BF: Membrane alterations in hemolysis: Internalizations of plasmalemma induced by primaquine. Science 164:843–845, 1969.

150. Thompson D: "On Growth and Form," abridged edition. Cambridge: Cambridge University Press, 1961, pp 157–158.

151. Chapman D: Phase transitions and fluidity characteristics of lipids and cell membranes. Q Rev Biophys 8:185–235, 1975.

152. Chapman D: Fluidity and phase transitions of cell membranes. In Eisenberg HK, Katchalski-Katzer R, Mauson LA (eds): "Biomembranes," Vol 7. New York: Plenum Press, 1975, pp 1–9.

153. Chapman D, Cornell BA: Phase transitions, protein aggregation and membrane fluidity. In Abramhamsson S, Pascher I (eds): "Structure of Biological Membranes." New York: Plenum Press, 1977, pp 85–93.

154. Mahoney EM, Hamill AL, Scott WA, Cohn ZA: Response of endocytosis to altered fatty acyl composition of macrophage phospholipids. Proc Natl Acad Sci USA 74:4895–4899, 1977.

155. Mahoney EM, Scott WA, Landsberger FR, Hamill AL, Cohn ZA: Influence of fatty acyl substitution on the composition and function of macrophage membranes. J Biol Chem 255:4910–4917, 1980.

156. Dunn WA, Hubbard AL, Aronson NN Jr: Low temperature selectively inhibits fusion between pinocytic vesicles and lysosomes during heterophagy of ^{125}I-asialofetuin by the perfused rat liver. J Biol Chem 255:5971–5978, 1980.

157. Kaplan J, Nielsen ML: Analysis of macrophage surface receptors. II. Internalization of α_2-macroglobulin trypsin complexes by rabbit alveolar macrophages. J Biol Chem 234:7329–7335, 1979.

158. Goldstein B, Wofsy C: Analysis of coated pit recycling on human fibroblasts. Cell Biophys 3:251–277, 1981.

159. Marsh M, Helenius A: Adsorptive endocytosis of Semliki forest virus. J Mol Biol 142:439–454, 1980.

160. Anderson RGW, Brown MS, Goldstein JL: Inefficient internalization of receptor-bound low density lipoprotein in human carcinoma A-431 cells. J Cell Biol 88:441–452, 1981.

161. Fisher GW, Rebhun LI: Turn on of endocytotic processes in response to sea urchin egg activation accompanies restructuring of the egg surface. J Cell Biol 91:185a, 1981.

162. Connolly JL, Green SA, Greene LA: Pit formation and rapid changes in surface morphology of sympathetic neurons in response to nerve growth factor. J Cell Biol 90:176–180, 1981.

163. Connolly JL: Personal communication.

164. Berlin RD, Oliver JM, Walter RJ: Surface functions during mitosis. I. phagocytosis, pinocytosis and mobility of surface-bound conA. Cell 15:327–341, 1978.
165. Albertini DF, Berlin RD, Oliver JM: The mechanism of concanavalin A cap formation in leukocytes. J Cell Sci 26:57–75, 1977.
166. Salisbury JL, Condeelis JS, Satir P: Role of coated vesicles, microfilaments, and calmodulin in receptor mediated endocytosis by cultured B lymphoblastoid cells. J Cell Biol 87:132–141, 1980.
167. Linden CD, Dedman JR, Chafouleas JG, Means AR, Roth TF: Interactions of calmodulin with coated vesicles from brain. Proc Natl Acad Sci USA 78:308–312, 1981.
168. Salisbury JL, Condeelis JS, Maihle NJ, Satir P: Calmodulin localization during capping and receptor-mediated endocytosis. Nature 294:163–166, 1981.
169. Pfeiffer JR, Oliver JM, Berlin RD: Topographical distribution of coated pits. Nature 286:727–729, 1980.
170. Walter R, Berlin RD, Pfeiffer JR, Oliver JM: Polarization of endocytosis and receptor topography on cultured macrophages. J Cell Biol 86:199–211, 1980.
171. McKeon FD, Reichardt LF, Heuser JE: Evidence for clathrin's role in the patching and capping of cell surface IgM. J Cell Biol 87:93a, 1980.
172. Sattilaro RF, LeCluyse E, Dentler WL: Associations between microtubules and coated vesicles in vitro. J Cell Biol 87:250a, 1980.
173. Fearon DT, Kaneko I, Thomson GG: Membrane distribution and adsorptive endocytosis by C3b receptors on human polymorphonuclear leukocytes. J Exp Med 153:1615–1628, 1981.
174. Goldstein JL, Brown MS, Stone NJ: Genetics of the LDL receptor: Evidence that the mutations affecting binding and internalization are allelic. Cell 12:629–641, 1977.
175. Brown MS, Goldstein JL: Analysis of a mutant strain of human fibroblasts with a defect in the internalization of receptor-bound low density lipoprotein. Cell 9:663–674, 1976.
176. Anderson RGW, Goldstein JL, Brown MS: A mutation that impairs the ability of lipoprotein receptors to localize in coated pits on the cell surface of human fibroblasts. Nature 270:695–699, 1977.
177. Ash JF, Louvard D, Singer SJ: Antibody-induced linkages of plasma membrane proteins to intracellular octomyosin-containing filaments in cultured fibroblasts. Proc Natl Acad Sci USA 74:5584–5588, 1977.
178. Huet C, Ash JF, Singer SJ: The antibody-induced clustering and endocytosis of HLA antigens on cultured human fibroblasts. Cell 21:429–438, 1980.
179. Beisiegel U, Kita T, Anderson RGW, Schneider WJ, Brown MS, Goldstein JL: Immunologic cross-reactivity of the low density lipoprotein receptor from bovine adrenal cortex, human fibroblasts, canine liver and adrenal gland, and rat liver. J Biol Chem 256:4071–4078, 1981.
180. Bretscher MS, Thomson JN, Pearse BMF: Coated pits act as molecular filters. Proc Natl Acad Sci USA 77:4156–4159, 1980.
181. Goldstein B, Wofsy C, Bell G: Interactions of low density lipoprotein receptors with coated pits on human fibroblasts: Estimate of the forward rate constant and comparison with the diffusion limit. Proc Natl Acad Sci USA 78:5695–5698, 1981.
182. Barak LS, Webb WW: Molecular motion of individual LDL molecules on cell membranes. Biophys J 33:74a, 1981.
183. Bretscher MS: Directed lipid flow in cell membranes. Nature 260:21–23, 1976.
184. Pastan IH, Willingham MC: Journey to the center of the cell: Role of the receptosome. Science 214:504–509, 1981.
185. Willingham MC, Maxfield FR, Paston IH: α_2-Macroglobulin binding to the plasma membrane of cultured fibroblasts: Diffuse binding followed by clustering in coated regions. J Cell Biol 82:614–625, 1979.

186. Dickson RB, Willingham MC, Pastan I: α_2-Macroglobulin adsorbed to colloidal gold: A new probe in the study of receptor-mediated endocytosis. J Cell Biol 89:29–34, 1981.
187. Willingham MC, Maxfield FR, Pastan I: Receptor-mediated endocytosis of alpha$_2$-macroglobulin in cultured fibroblasts. J Histochem Cytochem 28:818–823, 1980.
188. Haigler HT, McKanna JA, Cohen S: Direct visualization of the binding and internalization of a ferritin conjugate of epidermal growth factor in human carcinoma cells A-431. J Cell Biol 81:382–395, 1979.
189. McKanna JA, Haigler HT, Cohen S: Hormone receptor topology and dynamics: Morphological analysis using ferritin-labelled epidermal growth factor. Proc Natl Acad Sci USA 76:5689–5693, 1979.
190. Folk JE: Transglutaminases. Ann Rev Biochem 49:517–531, 1980.
191. Maxfield FR, Willingham MC, Davies PJA, Pastan I: Amines inhibit the clustering of α_2-macroglobulin and EGF on the fibroblast cell surface. Nature 277:661–663, 1979.
192. Davies PJA, Davies DR, Levitzki A, Maxfield FR, Milhaud P, Willingham MC, Pastan I: Transglutaminase is essential in receptor-mediated endocytosis of α_2-macroglobulin and polypeptide hormones. Nature 283:162–167, 1980.
193. Levitzki A, Willingham M, Pastan I: Evidence for participation of transglutaminase in receptor-mediated endocytosis. Proc Natl Acad Sci USA 77:2706–2710, 1980.
194. Haigler HT, Willingham MC, Pastan I: Inhibitors of ^{125}I-epidermal growth factor internalization. Biochem Biophys Res Commun 94:630–637, 1980.
195. Haigler HT, Maxfield FR, Willingham MC, Pastan I: Dansylcadaverine inhibits internalization of ^{125}I-epidermal growth factor in BALB 3T3 cells. J Biol Chem 255:1239–1241, 1980.
196. Dickson RB, Willingham MC, Pastan I: Binding and internalization of ^{125}I-α_2-macroglobulin by cultured fibroblasts. J Biol Chem 256:3454–3459, 1981.
197. FitzGerald D, Morris RE, Saelinger CB: Receptor-mediated internalization of Pseudomonas toxin by mouse fibroblasts. Cell 21:867–873, 1980.
198. Chaung D-M: Inhibitors of transglutaminase prevent agonist-mediated internalization of β-adrenergic receptors. J Biol Chem 256:8291–8293, 1981.
199. Tucciarone LM, Lanclos KD: Evidence for the involvement of transglutaminase in the uptake of vitellogenin by Xenopus laevis oocytes. Biochem Biophys Res Commun 99:221–227, 1981.
200. Baldwin D Jr, Prince M, Marshall S, Davies P, Olefsky JM: Regulation of insulin receptors: Evidence for involvement of an endocytic internalization pathway. Proc Natl Acad Sci USA 77:5975–5978, 1980.
201. Yarden Y, Gabbay M, Schlessinger J: Primary amines do not prevent the endocytosis of epidermal growth factor into 3T3 fibroblasts. Biochim Biophys Acta 674:188–203, 1981.
202. King AC, Hernandez-Davis L, Cuatrecasas P: Lysomotropic amines cause intracellular accumulation of receptors for epidermal growth factor. Proc Natl Acad Sci USA 77:3283–3287, 1980.
203. Van Leuven F, Cassiman J-J, Van den Berghe H: Uptake and degradation of α_2-macroglobulin-protease complexes in human cells in culture. Exp Cell Res 117:273–282, 1978.
204. Van Leuven F, Cassiman J-J, Van den Berghe H: Demonstration of an α_2-macroglobulin receptor in human fibroblasts, absent in tumor-derived cell lines. J Biol Chem 254:5155–5160, 1979.
205. Kaplan J, Nielsen MN: Analysis of macrophage surface receptors. I. Binding of α-macroglobulin protease complexes by rabbit alveolar macrophages. J Biol Chem 254:7323–7328, 1979.

206. Kaplan J: Evidence for reutilization of surface receptors for α-macroglobulin-protease complexes in rabbit alveolar macrophages. Cell 19:197–205, 1980.

207. Van Leuven F, Cassiman J-J, Van den Berghe H: Primary amines inhibit recycling of α_2-M receptors in fibroblasts. Cell 20:37–43, 1980.

208. Kaplan J, Keogh EA: Analysis of the effect of amines on inhibition of receptor-mediated and fluid-phase pinocytosis in rabbit alveolar macrophages. Cell 24:925–932, 1981.

209. Maxfield FR, Schlessinger J, Shechter Y, Pastan I, Willingham MC: Collection of insulin, EGF and α_2-macroglobulin in the same patches on the surface of cultured fibroblasts and common internalization. Cell 14:805–810, 1978.

210. Gulyas BJ, Matsuura S, Chen H-C, Yuan LC, Hodgen GD: Visualization of binding and internalization of a horseradish peroxidase-hCG conjugate by monkey luteal cells. Biol Reprod 25:609–620, 1981.

211. Goldstein JL, Brown MS, Anderson RGW: The low-density pathway in human fibroblasts: Biochemical and ultrastructural correlations. In Brinkley BR, Porter KR (eds): "International Cell Biology 1976–1977." New York: Rockefeller University Press, 1977, pp 639–648.

212. Louvard D: Apical membrane aminopeptidase appears at site of cell-cell contact in cultured kidney epithelial cells. Proc Natl Acad Sci USA 77:4132–4136, 1980.

213. Moxon LA, Wild AE, Slade BS: Localization of proteins in coated micropinocytotic vesicles during transport across rabbit yolk sac endoderm. Cell Tissue Res 171:175–193, 1976.

214. Edelson PJ, Cohen ZA: Effects of concanavalin A on mouse peritoneal macrophages. I. Stimulation of endocytic activity and inhibition of phago-lysosome formation. J Exp Med 140:1364–1386, 1974.

215. Bartholeyns J, Quintart J, Baudhuin P: Inhibition of the discharge of endocytosed protein from phagosomes into lysosomes in hepatoma cells exposed to dimerized ribonuclease A. Biochem J 178:433–442, 1979.

216. Goren MB, Hart PD, Young MR, Armstrong JA: Prevention of phagosome-lysosome fusion in cultured macrophages by sulfatides of Mycobacterium tuberculosis. Proc Natl Acad Sci USA 73:2510–2514, 1976.

217. Jones TC, Hirsch JG: The interaction between Toxoplasma gondii and mammalian cells. II. The absence of lysosomal fusion with phagocytic vacuoles containing living parasites. J Exp Med 136:1173–1194, 1972.

218. Ottosen PD, Courtoy PJ, Farquhar MG: Pathways followed by membrane recovered from the surface of plasma cells and myeloma cells. J Exp Med 152:1–19, 1980.

219. Farquhar MG: Recovery of surface membrane in anterior pituitary cells: Variations in traffic detected with anionic and cationic ferritin. J Cell Biol 77:R35–R42, 1978.

220. Maxfield FR, Tycko B: Rapid acidification of endocytic vesicles containing α_2-macroglobulin. J Cell Biol 91:212a, 1981.

221. Schneider Y-J, Tulkens P, deDuve C, Trouet A: Fate of plasma membrane during endocytosis. I. Uptake and processing of anti-plasma membrane and control immunoglobulins by cultured fibroblasts. J Cell Biol 82:449–465, 1979.

222. Schneider Y-J, Tulkens P, deDuve C, Trouet A: Fate of plasma membrane during endocytosis. II. Evidence for recycling (shuttle) of plasma membrane constituents. J Cell Biol 82:466–474, 1979.

223. Baumann H, Doyle D: Metabolic fate of cell surface glycoproteins during immunoglobulin-induced internalization. Cell 21:897–907, 1980.

224. Storrie B, Dreesen TD, Maurey KM: Rapid cell surface appearance of endocytic membrane proteins in chinese hamster ovary cells. Mol Cell Biol 1:261–268, 1981.

225. Herzog V, Farquhar MG: Luminal membrane retrieved after exocytosis reaches most Golgi cisternae in secretory cells. Proc Natl Acad Sci USA 74:5073–5077, 1977.

226. Goud B, Antoine J-C, Gonatas NK, Stieber A, Avrameas S: A comparative study of fluid-phase and adsorptive endocytosis of horseradish peroxidase in lymphoid cells. Exp Cell Res 132:375–386, 1981.
227. Herzog V, Reggio H: Pathways of endocytosis from luminal plasma membrane in rat exocrine pancreas. Eur J Cell Biol 21:141–150, 1980.
228. Van Deurs B, Von Bulow F, Moller M: Vesicular transport of cationized ferritin by the epithelium of the rat choroid plexus. J Cell Biol 89:131–139, 1981.
229. Pelletier G: Secretion and uptake of peroxidase by rat adenohypophyseal cells. J Ultrastruct Res 43:445–459, 1973.
230. Brown, MS, Anderson RGW, Basu SK, Goldstein JL: Recycling of cell-surface receptors: Observations from the LDL receptor system. Cold Spring Harbor Symposia 46:713–721, 1982.
231. Basu SK, Goldstein JL, Anderson RGW, Brown MS: Monensin interrupts the recycling of low density lipoprotein receptors in human fibroblasts. Cell 24:493–502, 1981.
232. Tanabe T, Pricer WE Jr, Ashwell G: Subcellular membrane topology and turnover of a rat hepatic binding protein specific for asialoglycoproteins. J Biol Chem 254:1038–1043, 1979.
233. Blumenthal R, Klausner RD, Weinstein JN: Voltage-dependent translocation of the asialoglycoprotein receptor across lipid membranes. Nature 288:333–338, 1980.
234. Stockert RJ, Howard DJ, Morell AG, Scheinberg IH: Functional segregation of hepatic receptors for asialoglycoproteins during endocytosis. J Biol Chem 255:9028–9029, 1980.
235. Stockert RJ, Gartner U, Morell AG, Wolkoff AW: Effects of receptor-specific antibody on the uptake of desialylated glycoproteins in the isolated perfused rat liver. J Biol Chem 255:3830–3831, 1980.
236. Doyle D, Hou E, Warren R: Transfer of the hepatocyte receptor for serum asialoglycoproteins to the plasma membrane of a fibroblast. J Biol Chem 254:6853–6856, 1979.
237. Marshall S, Green A, Olefsky JM: Evidence for recycling of insulin receptors in isolated rat adipocytes. J Biol Chem 256:11464–11470, 1981.
238. Weigel PH: Evidence that the hepatic asialoglycoprotein receptor is internalized during endocytosis and that receptor recycling can be uncoupled from endocytosis at low temperature. Biochem Biophys Res Commun 101:1419–1425, 1981.
239. Steer CJ, Ashwell G: Studies on a mammalian hepatic binding protein specific for asialoglycoproteins: Evidence for receptor recycling in isolated rat hepatocytes. J Biol Chem 255:3008–3013, 1980.
240. Tolleshaug H, Berg T: Chloroquine reduces the number of asialoglycoprotein receptors in the hepatocyte plasma membrane. Biochem Pharm 28:2919–2922, 1979.
241. Tietze C, Schlesinger P, Stahl P: Chloroquine and ammonium ion inhibit receptor-mediated endocytosis of mannose-glycoconjugates by macrophages: Apparent inhibition of receptor recycling. Biochem Biophys Res Commun 93:1–8, 1980.
242. Stahl P, Schlesinger PH, Sigardson E, Rodman JS, Lee YC: Receptor-mediated pinocytosis of mannose glycoconjugates by macrophages: Characterization and evidence for receptor recycling. Cell 19:207–215, 1980.
243. Gonzalez-Noriega A, Grubb JH, Talkad V, Sly WS: Chloroquine inhibits lysosomal enzyme pinocytosis and enhances lysosomal enzyme secretion by impairing receptor recycling. J Cell Biol 85:839–852, 1980.
244. Thyberg J, Nilsson J, Hellgren D: Recirculation of cationized ferritin in cultured mouse peritoneal macrophages: Electron microscopic and cytochemical studies with double-labelling technique. Eur J Cell Biol 23:85–94, 1980.
245. Muller WA, Steinman RM, Cohn ZA: The membrane proteins of the vacuolar system. I. Analysis by a novel method of intralysosomal iodination. J Cell Biol 86:292–303, 1980.

246. Muller WA, Steinman RM, Cohn ZA: The membrane proteins of the vacuolar system. II. Bidirectional flow between secondary lysosomes and plasma membrane. J Cell Biol 86:304–314, 1980.

247. Morris B, Morris R: Quantitative assessment of the transmission of labelled protein by the proximal and distal regions of the small intestine of young rats. J Physiol 255:619–634, 1976.

248. Geisow MJ, Hart PD, Young MR: Temporal changes in lysosome and phagosome pH during phagolysosome formation in macrophages: Studies by fluorescence spectroscopy. J Cell Biol 89:645–652, 1981.

249. Creutz CE, Pazoles CJ, Pollard HB: Identification and purification of an adrenal medullary protein (Synexin) that causes calcium-dependent aggregation of isolated chromaffin granules. J Biol Chem 253:2858–2866, 1978.

250. Pollard HB, Pazoles CJ, Creutz CE, Zinder O: The chromaffin granule and possible mechanisms of exocytosis. Int Rev Cytol 58:160–198, 1979.

251. Creutz CE: cis-Unsaturated fatty acids induce the fusion of chromaffin granules aggregated by Synexin. J Cell Biol 91:247–256, 1981.

252. Haigler H, Ash JF, Singer SJ, Cohen S: Visualization by fluorescence of the binding and internalization of epidermal growth factor in human carcinoma cells A-431. Proc Natl Acad Sci USA 75:3317–3321, 1978.

253. Gorden P, Carpentier J-L, Cohen S, Orci L: Epidermal growth factor: Morphological demonstration of binding, internalization, and lysosomal association in human fibroblasts. Proc Natl Acad Sci USA 75:5025–5029, 1978.

254. Soman V: Regulation of the glucagon receptor by physiological hyperglucagonaemia. Nature (Lond) 272:829–832, 1978.

255. Kasuga M, Kahn CR, Hedo JA, Van Obberghen E, Yamada KM: Insulin-induced receptor loss in cultured human lymphocytes is due to accelerated receptor degradation. Proc Natl Acad Sci USA 78:6917–6921, 1981.

256. Kahn CR, Baird K, Flier JS, Jarrot DB: Effects of autoantibodies to the insulin receptor on isolated adipocytes: Studies of insulin binding and insulin action. J Clin Invest 60:1094–1106, 1977.

257. Carpentier J-L, Van Obberghen E, Gorden P, Orci L: Surface distribution of ^{125}I-insulin in cultured human lympocytes. J Cell Biol 91:17–25, 1981.

258. Jarret L, Smith RM: Ultrastructural localization of insulin receptors on adipocytes. Proc Natl Acad Sci USA 72:3526–3530, 1975.

259. Gorden P, Carpentier J-L, Van Obberghen E, Barazzone P, Roth J, Orci L: Insulin-induced receptor loss in the cultured human lymphocyte: Quantitative morphological perturbations in the cell and plasma membrane. J Cell Sci 39:77–88, 1979.

260. Hinkle PM, Tashjian AH: Thyroid releasing hormone regulates the number of its own receptors in the GH^3 strain of pituitary cells in culture. Biochemistry 14:3845–3851, 1975.

261. Schonbrunn A, Tashjian AH Jr: Modulation of somatostatin receptors by thyrotropin-releasing hormone in a clonal pituitary cell strain. J Biol Chem 255:190–198, 1980.

262. Showell HJ, Williams D, Becker EL, Naccache PH, Sha'afi R: Desensitization and deactivation of the secretory responsiveness of rabbit neutrophils induced by the chemotactic peptide, formyl-methionyl-leucyl-phenylalanine. J Reticuloendothial Soc 25:139–150, 1979.

263. Niedel JE, Kahane I, Cuatrecasas P: Receptor-mediated internalization of fluorescent chemotactic peptide by human neutrophils. Science 205:1412–1414, 1979.

264. Stahl PD, Rodman JS, Miller MJ, Schlesinger PH: Evidence for receptor-mediated binding of glycoproteins, glycoconjugates, and lysosomal glycosidases by alveolar macrophages. Proc Natl Acad Sci USA 75:1399–1403, 1978.

265. Sherman LA, Lee J: Specific binding of soluble fibrin to macrophages. J Exp Med 145:76–85, 1977.
266. Youngdahl-Turner P, Mellman IS, Allen RH, Rosenberg LE: Protein mediated vitamin uptake: Adsorptive endocytosis of the transcobalamin II–cobalamin complex by cultured human fibroblasts. Exp Cell Res 118:127–134, 1979.
267. Takahashi K, Tavassoli M, Jacobsen DW: Receptor binding and internalization of immobilized transcobalamin II by mouse leukaemia cells. Nature 288:713–715, 1980.
268. Sullivan AL, Grasso JA, Weintraub LR: Micropinocytosis of transferrin by developing red cells: An electron-microscopic study utilizing ferritin-conjugated transferrin and ferritin-conjugated antibodies to transferrin. Blood 47:133–143, 1976.
269. Parmley RT, Ostroy F, Gams RA, DuLucas L: Ferrocyanide staining of transferrin and ferritin conjugated antibody to transferrin. J Histochem Cytochem 27:681–685, 1979.
270. Pike LJ, Lefowitz RJ: Activation and desensitization of β-adrenergic receptor coupled GTPase and adenylate cyclase of frog and turkey erythrocyte membranes. J Biol Chem 255:6860–6867, 1980.
271. Metzger H: The cellular receptor for IgE. In Cuatrecasas P, Greaves MF (eds): "Receptors and Recognition." London: Chapman and Hall, 1977, Vol 4, Ser A, pp 74–102.
272. Ishizaka T, Kshizaka K: Biology of immunoglobulin E: Molecular basis of reaginic hypersensitivity. Prog Allergy 19:60, 1975.
273. Wallace RA, Jared DW: Protein incorporation by isolated amphibian oocytes: Specificity for vitellogenin incorporation. J Cell Biol 69:345–351, 1976.

Modern Cell Biology, 1:53–117

Structure and Function of the Adhesive Glycoprotein Fibronectin

Leo T. Furcht

From the Department of Laboratory Medicine and Pathology, University of
Minnesota, Minneapolis, Minnesota 55455

I. HISTORY

Fibronectin is a high molecular weight glycoprotein present in plasma, cell matrices, basal lamina, and on cell surfaces. This molecule has been studied and known to investigators by various names. The first area of its investigation was as a plasma constituent where it was studied by Morrison et al in 1948 [1]. These studies dealt with plasma fractionation and coagulation factors and described the existence of a cold insoluble globulin. The second area of investigation was in relationship to the appearance of fibronectin and collagen in cell matrices and cell adhesion. The third major area of study was in reticuloendothelial function where fibronectin was known as α_2 opsonic protein. A number of reviews have appeared which deal with various aspects of fibronectin [2–9, and see others cited in specific sections].

Plasma fibronectin was first discovered in 1948 by Morrison et al [1], when a fibrinogen-rich plasma fraction was found to be cold-insoluble. Of further significance was the inability of this cold-precipitable fraction to be acted upon by thrombin [10]; however, later studies (see below) reported proteolytic digestion of fibronectin by thrombin. This protein was termed cold-insoluble globulin. Not much work was done on the plasma form of fibronectin until the studies of Edsall et al [10], who pursued plasma fractionation studies and Smith and Von Korff [11] and Mosesson and colleagues [12], who discovered a patient with a coagulopathy and cryofibrinogenemia secondary to a neoplasm. The cold-precipitable fraction from this patient consisted of fibrinogen and cold-insoluble globulin or, as it is now known, plasma fibronectin. Mosesson and associates showed that normal and diseased plasma have a fraction which precipitates with heparin or other mucopolysaccharides that consists of fibronectin [12, 13].

II. GENERAL PROPERTIES

The plasma form of fibronectin consists of a disulfide-bonded dimer with a molecular weight of 450,000 daltons and subunit chains of $\sim$ 220,000 daltons seen under reducing conditions [14, 15]. Studies by Alexander et al [16], and Mosesson et al [14] defined many of the physical chemical characteristics of fibronectin. Ultracentrifugation studies showed that fibronectin has an S value of about 12–14 in PBS, consistent with a molecular weight of 450,000. Sedimentation in 7 M urea without reductants gave a major peak at 5.6 S. Denaturing fibronectin in 7 M urea and reducing with dithiotheitol gave a peak at 3.9 S, suggesting that the original molecule was disulfide-bonded. Circular dichroism spectra indicate no good fit with α helix or β pleated sheet, and the molecule is therefore presumed to have globular and flexible regions. Physical chemical studies show that fibronectin begins to denature at 50–56°C and is complete at 70°C [16]. The cell matrix form of fibronectin is much less soluble than plasma fibronectin and requires rather harsh extraction conditions or conditions to maintain solubility such as 3 M urea and pH 11 [17]. Heparin has recently been shown to facilitate the extraction of cellular fibronectin, and the association of fibronectin with other matrix constituents is discussed below [18].

The two chains of the plasma fibronectin dimer appear to be identical by peptide mapping, though proteolysis of the presumed identical chains showed some differences in degradation rate of one chain over another [19]. Additionally, because of an apparent differential susceptibility to proteolytic cleavage and appearance of cleavage products, Fakuda et al [20] and Kurkinen et al [19] have suggested differences in the dimeric chains of plasma fibronectin which have been termed α and β [21]. Notwithstanding this contention, studies to date using peptide mapping, amino acid sequencing, or hybridoma antibodies have not delineated any unequivocal differences between the plasma fibronectin dimeric chains. Furthermore, the amino-terminal residue is blocked with pyroglutamic acid and appears to show an identical sequence of the first series of amino acids for the dimer [22]. This does not, however, rule out the possibility that there is a change at the carboxy terminal end of the more rapidly migrating band of the dimer. Whether some other modification of the chains may occur that contributes to the differences in chain migration, such as phosphorylation or glycosylation (see below), or whether separate genes coding for each of the chains might exist remains to be elucidated.

The plasma form of fibronectin migrates as a closely spaced doublet on reduced SDS polyacrylamide gels whereas the cellular form is a single band [14, 23–28]. Plasma fibronectin was initially purified by cold precipitation and ion-exchange chromatography [13]. Subsequent studies by Engvall and Ruoslahti [29] showed that gelatin-affinity chromatography could be used to purify plasma fibronectin, and this was rapidly adopted by many laboratories as a quick method to isolate fibronectin [30–32]. Most often fibronectin is eluted from gelatin columns with urea [29], but other methods such as sodium bromide, potassium iodide [30], ethylene glycol [31], amino compounds such as arginine [32], and lithium iodosalicylic acid [33] have been used. There has been one report suggesting that denaturing conditions such as urea will result in fibronectin preparations that are not opsonically active [34]. It remains to be seen whether other functions of fibronectin are perturbed by purification procedures.

Fibronectin is present in a number of body fluids, in addition to plasma, and one of the initial body fluids it was discovered in was amniotic fluid [24, 35–37]. Fibronectin is also present in cerebrospinal fluid at levels of 2–3 μg/ml [38], ascites fluid (unpublished observation), joint fluid [39], and urine [40] (discussed further below).

III. LOCALIZATION AND DISTRIBUTION

In general, fibronectin has relatively widespread distribution in tissues. Fibronectin is present in most loose connective tissue that has been examined, and is seen in basal lamina and associated with basement membranes [41–43; and see reviews 2–9]. Studies using light microscopy and immunofluorescence or immunoperoxidase showed the occurrence of fibronectin in virtually all loose connective tissue and basement membranes ranging from mammals to lower vertebrates such as the newt [41, 43, 44].

The studies on the localization of fibronectin in and around basement membranes, particularly glomerular basement membranes, have been somewhat conflicting. Fibronectin has been localized in the glomerulus, in the glomerular loops, basement membrane, and mesangial matrix in studies using immunofluorescence or light-microscopic immunoperoxidase [41, 43, 45–52]. In contrast to most reports, the earliest studies by Stennman and Vaheri [43] failed to observe fibronectin in the glomerular basement membrane; however, this may have been due to differences in fixation or processing. The studies of Pettersson and Colvin [46] reported that glomerular basement membrane fibronectin had very unique characteristics in comparison to other cell-derived fibronectin. For example, glomerular basement membrane fibronectin could not be released with reagents that generally can be used to extract fibronectin from cell matrices including 1) 8 M urea, 2) reducing agents, 3) nonionic detergents, 4) chaotropic agents, or 5) calcium chelation. It appears that fibronectin was covalently linked with basement membrane constituents but could be released by papain [46]. Electronmicroscopic studies provided more information on fibronectin organization within glomerular basement membrane [50, 52]. Studies by Madri et al [52] used ultrathin frozen sections and reported the coincident distribution of laminin antibodies along the glomerular basement membrane as had been shown for types IV and V collagen. Laminin was found only on the endothelial side of the capillary loops in the glomerular basement membrane comprising the lamina rara interna and lamina densa [52]. These same studies reported that fibronectin was absent from the glomerular basement membrane, but as others had shown at the light microscopic level, fibronectin was present in the mesangium. In contrast to this, Oberley et al [48] reported the ultrastructural localization of fibronectin on the endothelial and epithelial surfaces of the guinea pig glomerular basement membrane. In well-controlled studies using fixed or unfixed sections, peroxidase and ferritin procedures [50], other laboratories have observed findings similar to those reported by Oberley et al [48]. The studies of Courtoy et al [50] reported differing results depending on which methodology was utilized. Specifically, immunoperoxidase showed fibronectin to be on endothelial and epithelial surfaces, the laminae rarae internae and externae of the glomerular basement membrane. In the same studies, immunoferritin experiments showed fibronectin to be localized to subendothelial areas between endothelial and mesangial cells, in addition to the lamina rara interna. Fibronectin was not detected beneath the epithelial foot processes which was interpreted to be due to problems of penetration [50]. The precise distribution of fibronectin in the glomerular basement membrane will undoubtedly continue to be an area of active investigation. Finally, the presence of fibronectin in areas such as the basement membrane may be related to the molecular interactions fibronectin has with other matrix constituents known to be in basement membrane, such as proteoglycans [53].

Fibronectin has been shown to be present in skin of lower vertebrates, mammals, and man. In human tissue, fibronectin is observed in areas with a fair degree of collagen such as the dermal-epidermal junction and the dermis [54]. Fibronectin is largely in the basement membrane, papillary dermis, and surrounding epidermal appendages such as sweat glands and hair follicles [54]. This same distribution in skin has been observed in lower animals such as the newt [44].

Fibronectin has been shown to be synthesized by numerous cells in culture. Fibronectin production seems more predominant and undergoes various structural rearrangements in matrices of cells of mesodermal origin, but other cell types such as epithelial cells may also synthesize fibronectin. Among others, some of the cell types that synthesize fibronectin include the following: fibroblasts (see reviews and references therein); myoblasts [55–57]; chondrocytes [30, 58]; nervous system cells including astroglial cells grown in vitro [59] but not in vivo [60]; Schwann and Schwannoma cells [61] (Palm and Furcht, unpublished); melanoma cells [62]; endothelial cells [23, 63–66]; corneal endothelial cells [67]; glomerular cells [48, 68, 69]; various epithelial cells including what perhaps is the primary source of plasma fibronectin, hepatocytes [70]; renal and liver epithelial cells [71], intestinal epithelia [42]; breast and various other epithelial cells [72]. Fibronectin is clearly present in blood cells. For example, fibronectin is within platelets and may play a role in platelet aggregation [73–79]. Fibronectin is also synthesized by monocytes or macrophages which also have receptors for fibronectin [81–83]. It is also of note that a collagen-binding protein reacting with fibronectin antibodies has been observed on the surface of ejaculated sperm which 1) was removed by trypsin, 2) was recovered by incubation in seminal fluid, and 3) appeared to decrease with sperm capacitation [84].

IV. CARBOHYDRATES AND OTHER MODIFICATIONS

Plasma and cellular fibronectin are approximately 5% carbohydrate by weight [14, 28, 85, 86]. The carbohydrate composition of plasma fibronectin is 1.2% sialic acid, 1.8% hexose, and 2.1% hexosamine [14]. Fibronectins from other body fluids have different degrees of glycosylation. In particular, amniotic fluid fibronectin has significantly more carbohydrate, 7–9.5% [24, 37], and migrates more slowly on SDS-polyacrylamide gels. The sugars and the linkages appear to differ somewhat between plasma and cellular fibronectin. The majority of oligosaccharide chains are seen with a terminal sialic acid linked to galactose, acetylglucosamine, galactose and mannose, etc [14, 32, 85–89] via asparagine residues. The glycosylation of fibronectin is blocked by tunicamycin treatment of cells [90]. Tunicamycin will block transfer of sugars through dolicol phosphate intermediates. Cells incubated with tunicamycin produce unglycosylated fibronectin, and this has a more rapid movement through SDS-PAGE gels [90]. Also, unglycosylated fibronectin is more rapidly degraded by proteases. In contrast to this, retinoid treatment of cells increases the mobility and decreases the apparent molecular weight of fibronectin presumably because of an alteration in glycosylation when cells are grown in vitro [91].

Various radioactive sugars will be incorporated into fibronectin including in order of rate of incorporation ^{3}H-glucosamine $>$ ^{3}H-glucose, $>$ ^{3}H-fructose, $>$ ^{3}H-galactose, and $>$ ^{3}H-mannose. The carbohydrate component is asymmetrically distributed along the molecule [21], with some domains being highly glycosylated and others being virtually without carbohydrate (see reviews).

Fibronectin was observed to be present in amniotic fluid by a number of laboratories and suggested to be locally produced [24, 35, 38]. It was evident that the apparent molecular weight of the amniotic fluid fibronectin was slightly higher than that of plasma fibronectin [24, 36]. Studies by Ruoslahti [36] showed the amino acid composition and immunological reactions of amniotic and plasma fibronectin to be indistinguishable. Studies mentioned above reported the carbohydrate content of amniotic fluid fibronectin to be increased. It was stated that the peptide maps of plasma and amniotic fibronectins were similar, though in examining the published gels it does seem as if there might be differences [36]. This could represent problems in photography; however, the carbohydrates in amniotic fluid fibronectin were observed to be increased in fructose, glucosamine, galactosamine, and galactose [92], which is similar to some of the changes in the carbohydrates reported in cellular fibronectin [89]. Interestingly, the fibronectin produced by teratocarcinoma cells is somewhat similar to amniotic fluid fibronectin and has been proposed to be an oncodevelopmental marker of sorts [92].

These studies on inhibiting or enhancing glycosylation of fibronectin and comparing amniotic fluid and plasma fibronectin suggest a possible role for glycosylation in the apparent differences in chain molecular weights and molecular differences of fibronectins from varying sources. More detailed quantitative studies on the carbohydrates are necessary to characterize further what these differences may be.

In contrast to the enzymatic glycosylation of fibronectin that occurs at asparagine residues and can be blocked by tunicamycin, nonenzymatic glycosylation of fibronectin can occur (Chueng and Furcht, unpublished). Recently, we have observed that the plasma fibronectin in diabetic dogs is nonenzymatically glycosylated approximately 2.5–3 times that of normal dogs. Also, there is an excellent correlation between blood glucose and fibronectin nonenzymatic glycosylation. The fact that fibronectin is present in the glomerular mesangium and basement membrane (see below) points to the possible importance of this increased glycosylation of fibronectin in the microangiopathic changes that occur in diabetes mellitus.

In addition to these modifications that may occur on fibronectin, a variety of others exist. Disulfide bond formation leading to multimeric forms occurs and is discussed below. Covalent transglutamination reactions occur, and this is discussed in the section on Platelets and Coagulation. Fibronectin can also be covalently modified by phosphorylation. Teng and Rifkin [93] reported that normal and Rous sarcoma virus transformed chick embryo fibroblasts would incorporate ^{32}P-orthophosphate into fibronectin. The specific activity of ^{32}P-orthophosphate fibronectin was higher in normal than in transformed cells. ^{32}P was covalently linked to fibronectin and was found on phosphoserine and phosphothreonine after acid hydrolysis. It was estimated that there were two molecules of ^{32}P-phosphate incorporated per molecule of fibronectin. A number of other studies have shown that radioactive sulfate can be incorporated into fibronectin produced by cells in vitro [8, 62].

V. PLASMA AND CELLULAR FIBRONECTIN

Plasma and cellular fibronectin were first reported to be immunologically related based on the cross reaction of antibodies to both of the molecules [94, 95]. The most profound difference between plasma and cellular fibronectin is solubility. Cellular fibronectin requires harsh denaturing conditions to extract it and requires a very high pH—ie, 10–11—to maintain its solubility [17].

Most investigators have observed that matrix fibronectin, and to a greater extent amniotic fluid fibronectin (discussed above), is slightly retarded in migration in SDS gels giving a somewhat larger apparent molecular weight [96, 97]. However, some investigators have reported no difference in the apparent molecular weights between matrix and plasma fibronectins [28, 98–100], though others assert there are [100].

Additionally, plasma fibronectin migrates as a closely spaced doublet whereas cellular fibronectin migrates as a single somewhat broader apparently monomeric band. Glycosylation has always plagued molecular weight estimates obtained from SDS-PAGE. The total carbohydrate content for plasma and cellular has been reported to be similar [86–89]. However, cellular compared to plasma fibronectin is reported to contain less sialic acid [28, 88] and to have fructose. The amino acid compositions of plasma and cellular fibronectin are similar in a number of species that have been tested [14, 20, 30, 102–105].

The observation of Hayman and Ruoslahti [106] and Oh et al [107], showing that endogenous plasma fibronectin could exchange with matrix fibronectin, makes it a little confusing as to what truly represents cell matrix fibronectin. Perhaps cellular fibronectin represents a combination of endogenous, possibly unique cellular fibronectin which is highly insoluble, in addition to another component derived from plasma or tissue culture medium. A report by Atherton and Hynes [108] suggests a difference in plasma and cellular fibronectin based on differential affinity of monoclonal antibodies. Since a monoclonal antibody a priori recognizes a single epitope, one would generally postulate an all-or-none reaction of an antigen to a monoclonal antibody. A minimal conformational difference or amino acid substitution at the site of the purported difference where a monoclonal antibody binds could account for the differential affinity of monoclonal antibodies to plasma and cellular fibronectin reported [108]. Other studies show reaction of monoclonal antibody to plasma and cellular fibronectin. Based on the known profound differences in plasma and cellular fibronectin it would not be surprising to find conformational differences in the two, which might account for differential antibody affinity depending on the antibody binding site. If in fact there are true differences in plasma and cellular fibronectin that had an immunological basis, it would be more convincing to have monoclonal antibodies that reacted with one and not the other than to have monoclonal antibodies that reacted somewhat with both.

Most recently, studies from Yamada's laboratory have shown that there appears to be specific peptide differences in cellular and plasma fibronectin in the chicken [109]. This observation would suggest that separate genes may exist for plasma and cellular fibronectins because the reported differences occurred within the polypeptide chain, not at the amino or carboxy terminus, and therefore would not imply different processing or posttranslational modifications.

Fibronectin produced by cells in vitro has been reported to occur in three forms which may be unique or interchangeable; 1) an extracellular matrix form, 2) a diffuse plasma membrane form, and 3) a soluble form in supernatant medium. The matrix form has been documented extensively in various immunofluorescence studies (see reviews and references therein). The light microscopic studies from numerous laboratories showed that fibronectin occurs in a dense extracellular matrix most prominently at sites of cell-cell and cell-substrate interaction. Importantly, studies have shown that there is a codistribution of fibronectin and collagen in the cell matrix [110–114]. It would appear that fibronectin, collagen, and proteoglycans [115–117] all occur in this cell matrix or substrate-attached material (Figs. 1, 2).

Ultrastructural studies proved informative in detailing extracellular matrix constituents. Immunoelectron microscopic lectin cytochemistry studies added to the traditional electron microscopic studies. These studies showed a fibrillar cell matrix outside of myoblasts which reacted with concanavalin A [118]. This con A binding cell matrix disappeared as myoblasts differentiated (see differentiation section below).

The matrix form of fibronectin seen at the electron microscopic level in cells cultivated in the absence of ascorbate appears as nonperiodic, wispy fibrils approximately 6–10 nm in diameter, which coalesce to form bundles (Fig. 3). These 6- to 10-nm fibronectin filaments are most pronounced at sites of cell-cell and cell-substratum contact, and are juxtaposed with intracellular microfilaments (Fig. 4) [111–114, 119]. Antibodies to fibronectin and procollagen codistribute on these 6- to 10-nm fibrils at the ultrastructural level, which is consistent with immunofluorescence localization studies though these latter studies cannot resolve 6- to 10-nm fibrils [112–114]. These 6- to 10-nm fibrils are exquisitely sensitive to various proteases, including trypsin and plasmin, which is different from the effects of proteases on collagen fibrils formed in vivo. It is important to note that complete processing and synthesis of collagen by cells requires a number of cofactors. One cofactor, ascorbic acid, has only a very short half-life in vitro, on the order of several hours [120].

Ascorbic acid is a cofactor for prolyl hydroxylase, a rate-limiting enzyme in collagen biosynthesis. Daily addition of ascorbate to human fibroblasts will lead to the formation of larger diameter fibrils, approximately 40 nm in diameter [113, 114] (Fig. 5). Fibronectin is discontinuously localized with a 70-nm axial repeat along fibrils produced in the presence of ascorbate [113, 114] (Fig. 6).

The 70-nm axial repeat of fibronectin on these collagen fibrils is that distance separating sites on the collagen fiber that are susceptible to degradation by mammalian collagenase [121]. Preliminary studies show that fibronectin may protect somewhat the collagen fibril from degradation by collagenase (unpublished observation). Also, the larger diameter collagen and fibronectin fibrils produced by ascorbate treated cells are resistant to degradation by trypsin or related proteases (unpublished observation). These larger diameter fibrils seem more consistent with bona fide collagen fibrils

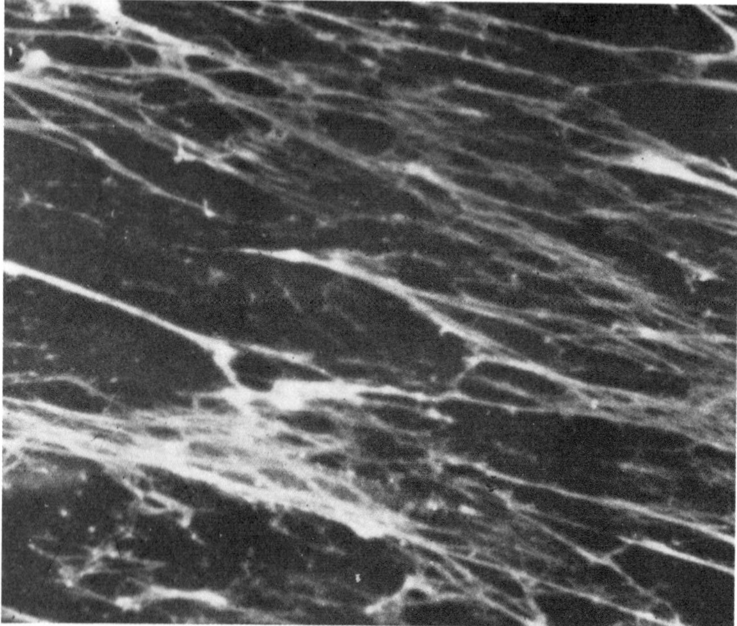

Fig. 1. Immunofluorescence localization of fibronectin antibodies on high-density human fibroblasts; note dense extracellular fibrils forming a matrix. (800 ×.)

seen in vivo, and also do not have the apparent contiguity with intracellular microfilaments like the smaller diameter fibrils seen without ascorbate treatment. Some cells appear able to process collagen in vitro more thoroughly even in the absence of ascorbate, but these seem to be the exceptions. One example is the 3T6 cell line which will produce large diameter fibrils in vitro, which seem very similar to collagen fibers seen in vivo [122].

Studies have raised the question whether the extracellular fibrils seen in cells cultured in vitro are representative of "collagen" fibrils seen in connective tissue in vivo. The thin nonperiodic 6- to 10-nm filaments can be seen in vivo in the embryo [123], occur in granulation tissue and upon migration of cells into wounds. The larger diameter fibrils formed in vitro in the presence of ascorbate are somewhat more typical of fibrils seen in vivo. It appears that both types of fibrils, the 6- to 10-nm and the thicker 40-nm fibril with fibronectin periodicity are both forms of collagen/fibronectin fibrils that can occur in vivo. One potential concern relative to defining cellular and plasma fibronectin is the important observation by Hayman and Ruoslahti [106] discussed above, that exogenous plasma fibronectin can be

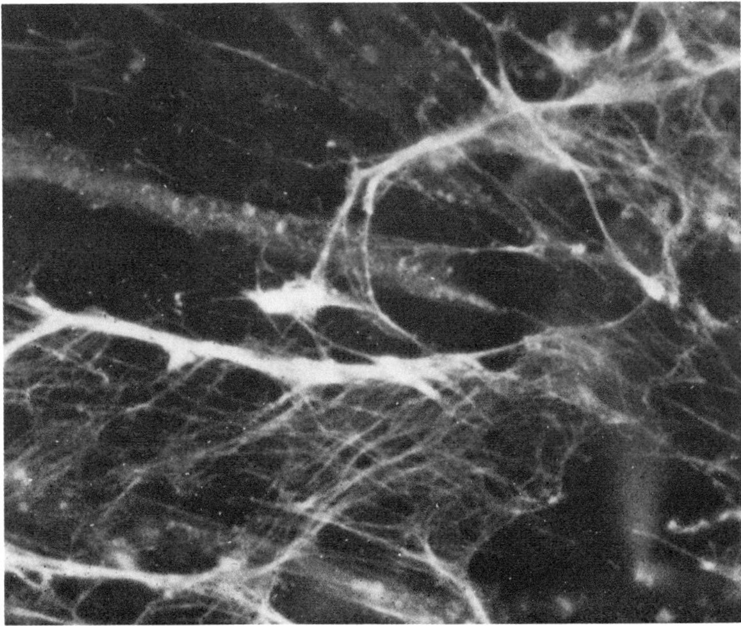

Fig. 2. Immunofluorescence localization of procollagen type I antibodies on high-density human fibroblasts; there is similar reactivity to antifibronectin seen alone. (800 ×.)

incorporated into cell matrices. These data were generated by using species-specific polyclonal antibodies, and we have made similar observations using monoclonal antibodies (unpublished). It has not been shown, however, that the plasma fibronectin added to cells, which is incorporated into the matrix, has the same insolubility characteristics and so on of true "cellular" fibronectin. The plasma fibronectin that is incorporated into matrices in the experiments described above may only be loosely associated to the matrix collagen or proteoglycan produced by the cell. Alternatively, the plasma fibronectin added to cells and incorporated into matrices may undergo modifications that affect its behavior so that it is biochemically like "cellular" fibronectin.

In addition to the matrix form of fibronectin, ultrastructural studies have shown fibronectin to be distributed along the plasma membrane in a diffuse fashion [112–114]. Another report showed globular, nonfibrillar accumulations of fibronectin near the plasma membrane [111]. Other investigators were unable to detect plasma membrane fibronectin (see review [3]). However, in support of the existence of plasma membrane fibronectin, antibodies to fibronectin are cytotoxic to cells [124].

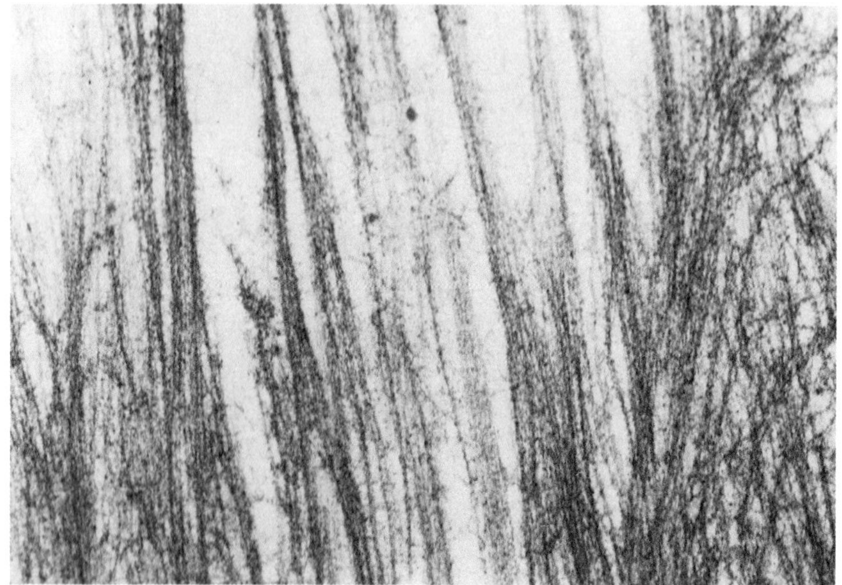

Fig. 3.   Electron micrograph of extracellular fibrils produced in vitro in the absence of ascorbate by human fibroblasts. Fibrils occur singly and coalesce into bundles and are 6–10 nm in diameter. (48,000 ×.)

VI. CELL ATTACHMENT AND SPREADING

Cell adhesion is a very complex and biologically important phenomenon. This process can clearly be related to many serum components, surface components, ions, and cytoskeletal structures [see 9, 125]. The review of Grinnell [125] is particularly useful in describing the whole physiology and requirements for cell adhesion. Among various surface components that may be responsible, the noncollagen glycoproteins fibronectin and laminin have attracted a great deal of interest. Since these glycoproteins interact with various components including proteoglycans and collagens, the production and type of these matrix constituents can be crucial in dictating or modifying cell adhesion. Some years ago, studies in tissue culture prior to the development of certain plastics showed that collagen substrata would facilitate the growth of many cell types in glass dishes. In fact, it was virtually obligatory to use collagen substrata to get growth of certain cells [126], and collagen could significantly affect the differentiation of many cell types, notably myoblasts [127].

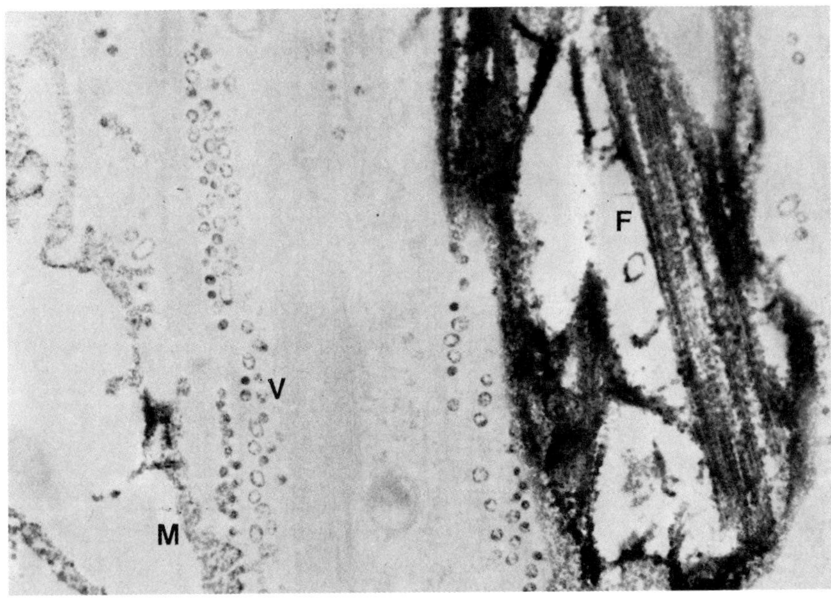

Fig. 4. Immunoperoxidase localization of fibronectin on the outside of human fibroblasts. Fibronectin is seen in extracellular, nonperiodic fibrils (F), a diffuse membrane-associated form (M), and in plasma membrane-associated vesicles (V). (68,000 ×.)

Studies by Klebe first showed the presence of a serum factor that would promote the adhesion of cells to collagen [128]. This factor is now known as fibronectin. Fibronectin seems most important, though certainly not exclusive, in acting as an attachment factor for cells of mesodermal origin. The initial studies in myoblasts by Haushka and White [127] showed the necessity for a serum factor to promote adhesion of cells to the substrate, which was confirmed by others [100]. Subsequent studies of Klebe [128] showed the presence of a serum factor which promoted the adhesion of fibroblasts to collagen. There are some cases in which fibronectin seems to mediate adhesion of epithelial cells such as liver [129] or metastatic pulmonary tumors [130] and cells of unknown type such as 3T3 cells [see reviews 9, 125, 131, 132].

In general, laminin appears to be the primary protein used for cell attachment by epithelial cells, but fibronectin appears to be able to do this also [see 9]. Chrondronectin, a recently discovered serum and cell matrix protein from cartilage, acts as an adhesion and attachment factor for chondrocytes

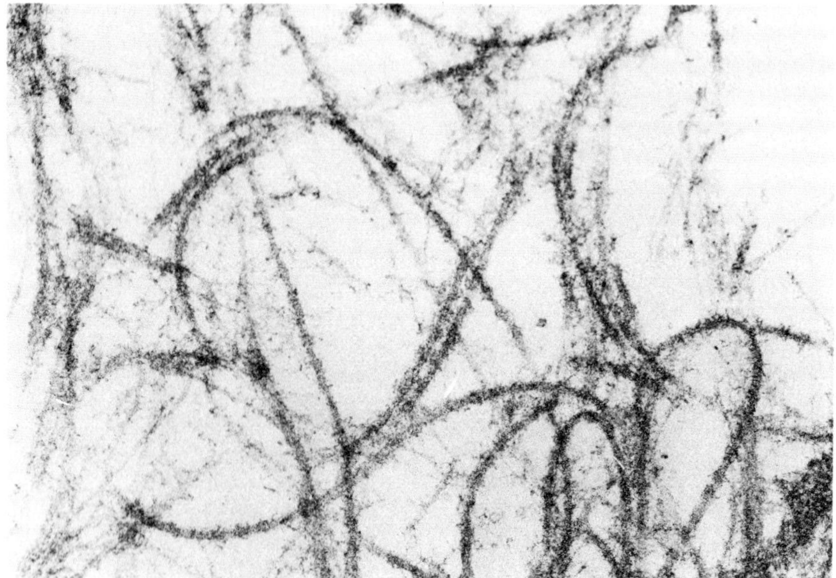

Fig. 5. Electron micrograph of extracellular fibrils produced by human fibroblasts treated in vitro with 50 μg/ml of ascorbic acid for 1 week every other day. Fibrils are curved, and larger in diameter (40 nm) than fibrils produced by cells grown in the absence of ascorbic acid. (83,000 ×.)

[133]. It is likely that there will be many newly defined cell attachment proteins which are in cell matrices and possibly also in plasma.

Interestingly, both plasma fibronectin and cell-extracted fibronectin are effective in promoting cell adhesion in serum-free media also [134]. There are substantial differences in the solubility of plasma and cellular fibronectin, as discussed above; however, both sources of fibronectin promote adhesion of cells to collagen at approximately 1 μg/ml. However, fibronectin from the cell layers of normal cells is 80 times more effective than plasma fibronectin in promoting a more normal-appearing phenotype when added to transformed cells [27, 135, 136]. It should be noted that cell-extracted fibronectin may carry along trace contaminants of other matrix constituents, perhaps proteoglycan such as heparan sulfate, that leads to the cell-extracted fibronectin being more efficacious than plasma fibronectin in restoring "normal" morphology to transformed cells. There appears to be a unique "domain," or site on the fibronectin molecule, which mediates cell attachment (see below).

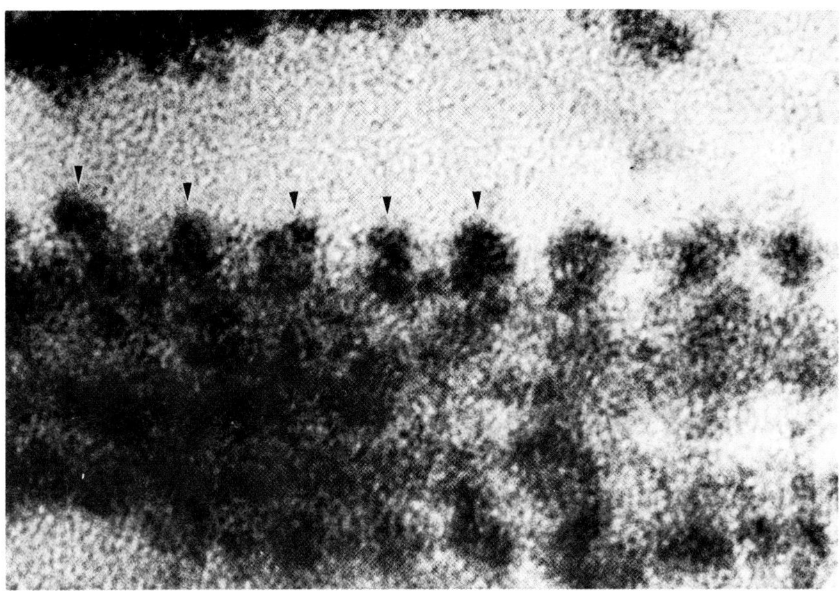

Fig. 6. Immunoperoxidase localization of fibronectin on fibrils formed by human fibroblasts in the presence of ascorbic acid. Fibronectin reacts with these larger-diameter fibrils in a periodic fashion with an axial repeat of 70 nm, indicated by the arrowheads. (140,000 ×.)

Studies by Grinnell and Minter [137] and Gold and Pearlstein [138], among others, evaluated the ability of chemically modified fibronectin to mediate cell attachment. An interesting observation was that reduced and alkylated fibronectin was still effective in promoting cell adhesion. This suggests that the monomeric fibronectin chain could promote adhesion, and it has been shown that peptide fragments could mediate cell adhesion. Succinylation of amino groups on fibronectin failed to modify its activity; however, in one case modification of carboxyl groups decreased cell adhesion activity [137], while Gold and Pearlstein [138] failed to observe the latter.

Studies on cell adhesion have often looked at cells that are capable of synthesizing molecules which have adhesive functions, such as fibronectin or laminin. An important control to perform is blocking the synthesis of these putative adhesion molecules with cyclohexamide [see 9, 125, and references therein]. In studies using WI-38 cells or other cells and examining fibronectin and intracellular actin, it was observed that focal attachment plaques did not contain fibronectin but did contain actin containing intracellular fibers [139, 140]. Some studies, however, have shown that fibronectin is localized at sites of "close contact" with the substratum rather than at focal adhesion sites [140, 141]. Other studies report that fibronectin may actively be removed from areas of focal contacts between cell and substratum [142].

VII. COLLAGEN-FIBRONECTIN INTERACTION

Fibronectin will bind all the known types of mammalian collagens. The affinity of fibronectin to denatured collagens is greater than for native collagens [143, 144]. Precisely why denatured collagen is more effective than native collagen in binding fibronectin is not known. There are specific binding sites in the various genetically distinct collagen chains that appear to interact with a region near the amino terminus of the fibronectin molecule (see below). The binding sites for fibronectin reside in the following cyanogen bromide fragments of collagen chains: $\alpha1(I)$-CB-7; $\alpha1(II)$ CB-10 [143–146]; $\alpha2$-CB-5; and a fragment of $\alpha1(III)$ [9]. The affinity of fibronectin for different collagens varies with the type. For native collagens, type III seemed to have greater affinity for fibronectin than type I [9, 143, 144]. A characteristic of the cyanogen bromide fragments of collagen reacting with fibronectin is that these constitute regions which are cleaved with mammalian collagenase. Interestingly, the interaction of collagens and fibronectin is modified by proteoglycan, namely heparin or heparan sulfate [147–150]; that is, the affinity of fibronectin and collagen for each other is enhanced by heparin addition. Also, proteoglycans form a major component of the extracellular matrix or substrate-attached material [115–117, 151]. This suggests that all these matrix constituents can interact with one another. Specific "domains" of fibronectin are responsible for these molecular interactions, and these are described below.

The molecular mechanism for the collagen-fibronectin interaction is not well known; however, some recent work has shed some light on this [32, 152]. It has been shown by some investigators that the denatured collagen-fibronectin interaction has a pH optimum at approximately 7–9 [152]; others report no effect of pH, ranging from 2.6 to 10.6 [153]. Increasing ionic strength of NaCl to even 0.1 M during the initial binding gave a 41% inhibition of binding, but once fibronectin and collagen interact, harsher conditions are required to dissociate them [152]. Succinylation of amino groups completely inhibited the fibronectin gelatin interaction, whereas perturbation of arginine residues, or carboxy groups with carbodi-imide, appears to modify charge of the fibronectin molecule and interaction with collagen [152]. Ionic detergents such as SDS and sodium deoxycholate inhibit fibronectin-collagen interaction [152, 153]. Nonionic detergents including Triton X-100, Tween 20, and Lubrol WX appear to have minimal effect on fibronectin-collagen interaction on collagen affinity columns [152, 154]. These observations are discordant with those of Gold and Pearlstein [153], who adsorbed collagen to plastic rather than covalently coupling it to an affinity matrix. Reduction and alkylation of fibronectin will block fibronectin from interacting with collagen [155]. There are convincing data to suggest that the predominant interaction between collagen and fibronectin is an ionic one, and this is supported by the observation that nonionic detergents that would perturb hydrophobic interactions have little effect [29, 152]. Finally, chemical modification studies point to lysine and arginine residues as being important in mediating these interactions [152].

VIII. CELL SURFACE RECEPTOR FOR FIBRONECTIN

It has not been unequivocally demonstrated what the precise surface membrane constituent is that binds fibronectin. Immunoelectron microscopic studies have clearly shown fibronectin to be a constituent on the surface of normal cells, and to a lesser extent on transformed cells [112–114]. Studies by Kleinman et al [156] showed that certain complex glycolipids—ie, the gangliosides GT_1 and GD_{1a}—could block the fibronectin-mediated adhesion of cells to a collagen substratum, suggesting that these constituents serve as "receptors" for surface fibronectin. It is also likely that glycoconjugates similar to these gangliosides could act as binding sites for fibronectin. It is important to note that the vast majority of fibronectin is not directly opposed to the plasma membrane, but is associated in dense fibrillar arrays away from the cell membrane (see various reviews and Fig. 4). It is also discussed in other sections that fibronectin interacts with other matrix constituents such as collagen or proteoglycan like heparan sulfate. Collagen or procollagen could serve as a "receptor," or be one of the binding components for fibronectin on the membrane of cells. The converse may also be true—that is, fibronectin may serve as a binding site for collagen or procollagen on the cell surface. In support of the potential role for surface-associated collagen or procollagen acting as a binding site for fibronectin is the observation that antibodies to collagen are cytotoxic to cells when added in vitro, which suggests that the collagen is a surface constituent [157–160]. Also, as discussed in detail in the section on platelets, antibodies to fibronectin will partially block platelet aggregation induced by collagen [75], and one report suggests that fibronectin is the platelet "receptor" for collagen [73]. Both of these situations with surface-associated fibronectin or collagen binding the other seems plausible. It is also of note that preliminary observations show that supernatant tissue culture medium of normal cells contains soluble fibronectin-collagen complexes which can be specifically immunoprecipitated with a monoclonal antibody to fibronectin (Smith and Furcht, unpublished). This could be interpreted to indicate that either molecule could potentially serve as a binding site for the other. In the same vein, recent studies from Hook's laboratory have shown that a percentage of heparan sulfate can be loosely associated with the plasma membrane, and that there is another heparan sulfate component that appears to be tightly associated, perhaps hydrophobically, with lipid, and is like an integral membrane component [161]. This heparan sulfate, which is tightly interacting with lipid, could very likely serve as one of the binding components for surface-associated fibronectin and thereby indirectly for collagen (see Fig. 9). This intramembranous heparan sulfate could also serve in providing a transmembranous link between the cytoskeleton and the matrix exoskeleton (see below). A number of reports from Culp's laboratory have proposed that heparan sulfate may serve as a receptor for surface fibronectin [115–117]. Further studies with immunoelectron microscopy or chemically cross-linking surface constituents will permit a more clear understanding of which components may bind to each other. Since proteoglycan, fibronectin, and collagen can all interact, it would not be surprising to see these all binding together or being chemically linked on the surface. Additionally, recent studies have shown that heparin will greatly facilitate the extraction of fibronectin from lung and other tissues [18]. It is also interesting to note that heparin is necessary to promote the fibronectin stimulated opsonization of gelatin particles by reticuloendothelial cells (see below). These various studies appear to indicate that any of the matrix constitutents could serve in binding the others on the cell surface. Also, these components may exist in a soluble complex which may bind or be covalently linked to other constituents.

It is also of note that various proteases will release fibronectin from cells, suggesting that protein is one of the classes of molecules acting to bind fibronectin. Some of these proteases will degrade fibronectin; others, such as thrombin, stimulate the release of apparently intact fibronectin [162]. There does appear to be some covalent association of fibronectin on or outside of cells by disulfide binding [163]. Also, transglutamination via factor XIII$_a$ could provide a mechanism for surface covalent associations [164, 165].

Clearly, the processes that lead to fibronectin binding and fibril formation are dynamic ones. For example, normal fibroblasts at low cell density have little matrix fibronectin, but upon contacting one another, at high cell density, matrix fibronectin forms [112]. Also, when cells have made matrix fibronectin, as cells progress from G to M, cells round up and release from this matrix fibronectin. Recent work has shown that addition of cell-derived fibronectin to cells promoted the formation of focal adhesions and could promote growth as evidenced by entry into S phase [166].

IX. FIBRONECTIN AND CYTOSKELETAL STRUCTURES

There appears to be a general correlation in many mesodermal cells of cell adhesion, surface-associated fibrillar fibronectin, and a high degree of cytoskeletal organization. A recent review of this specific area can be seen in Hynes [167]. Much of the work in this area grew out of studies on the cytoskeleton in normal and transformed cells. The cytoskeleton is composed of three main structural constituents; microfilaments, intermediate filaments, and microtubules [see 168 and references therein]. Contact inhibited cells in vitro have highly structured microfilaments which coalesce to form stress fibers or actin cables [169]. These microfilament bundles run to the plasma membrane and substrate adhesion sites, namely focal adhesion plaques. Microfilaments are 7 nm in diameter and label with antibodies to actin [see references in 167, 168]. Intracellular microfilaments appear to terminate with the plasma membrane and at dense bodies seen within focal attachment plaques (Fig. 7). Immunofluorescent studies have shown α-actinin, a Z

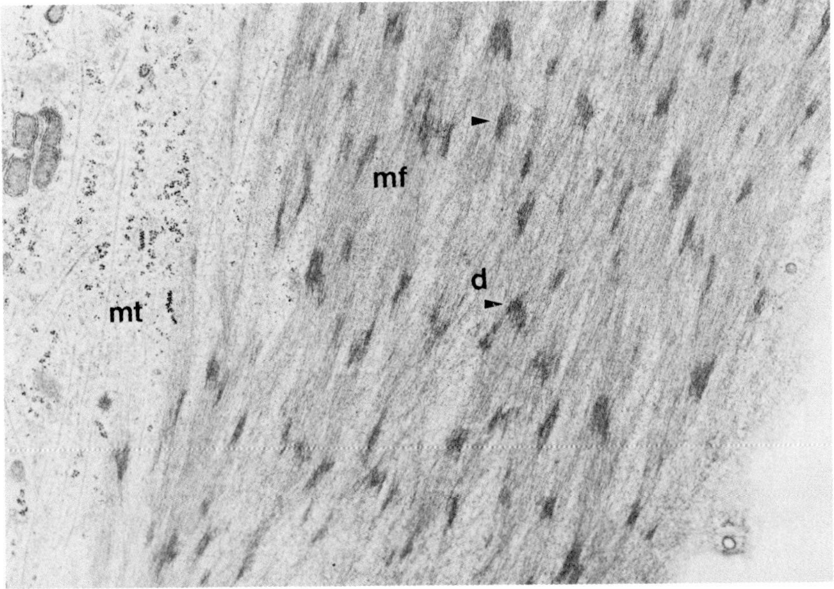

Fig. 7. Electron micrograph of human fibroblasts. Note dense bundles of microfilaments (MF) coursing through the periphery of the cytoplasm often coalescing at dense bodies (D, arrows). Microtubules (MT) are seen adjacent to this zone of microfilaments.

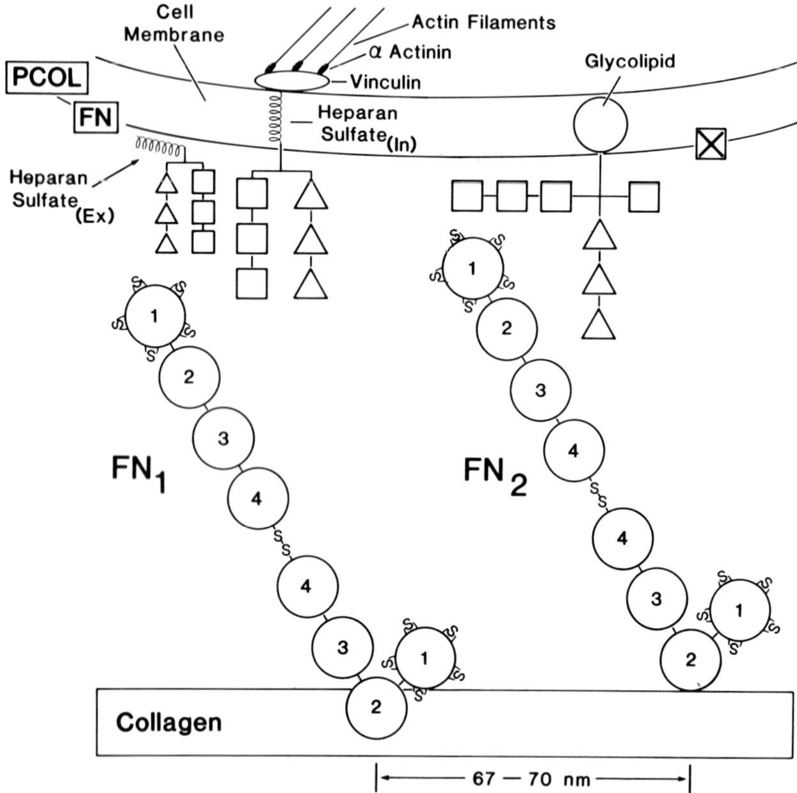

Fig. 8. Schematic representation of cell membrane/fibronectin/collagen/cytoskeleton interaction. Fibronectin is shown as globular "domains" disulfide linked to produce dimers. Four domains are shown, though others are known to exist (see Fig. 9) beginning at the amino terminus with domain 1. Domain 1 is shown with five disulfide loops, or fingers, that constitute a reiteration of an identical sequence which is repeated three times in a COOH terminal domain [350]. Domain 1 is also the weak heparin binding and the Staphylococcal binding region. Domain 2 is the gelatin binding region; domain 3 is the cell-cell binding domain and a strong heparin-binding region. This is further subdivided in Figure 9. The interchain disulfide bonds are at the COOH terminus in domain 4. The interaction with collagen fibrils is represented in FN$_1$ at what would be a covalent linkage via the interaction of the gelatin binding domain 2 with the collagen fibrils and cross-linking at domain 1. FN$_1$ is shown with a loose, noncovalent linkage of the gelatin-binding domain (2) with the collagen fibril. The intramolecular spacing of fibronectin on the collagen fibril is shown to be 67–70 nm by ultrastructural immunoperoxidase and immunoferritin studies [113, 114]. Potential membrane-binding sites for fibronectin include glycolipids and proteoglycan shown as heparan sulfate; the latter can be an

line–associated protein in striated muscle, to be near the peripheral termination of microfilaments [170–172]. Other studies have shown a 130,000-dalton protein immediately subjacent to the cytoplasmic leaflet of the plasmalemma, and this protein has been called vinculin [172] or focin [173]. Immunoelectron microscopic studies have shown α-actinin in "dense bodies" and near the plasmalemma though vinculin appears more closely associated with the plasma membrane than α-actinin [170, 172] (Fig. 7, and see model Fig. 8).

Microfilaments, separate from those involved with focal attachment or adhesion sites, are most pronounced at sites of cell-cell and cell-substrate attachment zones [169, 174, 175]. Microfilaments will increase at sites of cell-cell contact, and this parallels the development of extracellular matrix fibronectin; that is, as cells increase contact with one another, there is increased development of an "exoskeleton" of fibronectin and other matrix components [112, 118].

Transformed fibroblast cells are, in general, less adhesive, and have a rather disorganized cytoskeleton of microfilaments and microtubules compared to normal cells [174, 176, 177, review 167, 168]. Interestingly, when certain transformed cells are cloned and variants with increased adhesion are selected, these showed well-developed microfilament bundles [178]. This again points to some general association of a high degree of cell adhesion with a well-developed cytoskeleton.

Extracellular fibronectin is associated with enhanced cell adhesion, and promotion of cell adhesion may secondarily lead to microfilament bundle formation. In support of this, when exogenous fibronectin is added to transformed cells, the cells flatten out, become more adhesive, and develop a

---

intrinsic membrane component (heparan sulfate—In) or a loosely associated membrane component (heparan sulfate—Ex) [161]. The carbohydrates on proteoglycan shown as heparan sulfate are represented as triangles and squares; the binding of heparan sulfate Ex with the plasma membrane is due to carbohydrate interactions with membrane components [161]. Other potential binding or cross-linking sites on the cell surface for fibronectin include procollagen (PCOL) or collagen, fibronectin (FN), and (X)—presently unknown membrane constituents. There may be transmembraneous associations of the fibronectin, collagen, proteoglycan, "exoskeleton" to the cytoskeleton as shown for FN_1. This is shown with vinculin immediately subjacent to the cytoplasmic leaflet of the bylayer, and with actin microfilaments and alpha-actinin terminating at vinculin. The specific molecular association of a surface "receptor" component with vinculin or the cytoskeleton has not been shown, but intermediary components may exist. There are clearly also situations where fibronectin is bound to the cell surface without juxtaposition to the cytoskeleton as shown for FN_2; the potential surface-binding site for FN in these cases would not always have to be glycolipids as shown, but could be fibronectin, procollagen, or other components.

more structured cytoskeleton [179, 180]. Also, when normal cells go into the mitotic phase of the cycle, cells round up, are less adhesive and detach from the fibronectin-collagen matrix, and have a disorganized microfilament system [167, 168]. Other studies have shown that proteases mimic what occurs during mitosis, and have a parallel effect on the cell matrix and cytoskeleton; that is, proteolysis will release surface fibronectin matrix fibrils and will disrupt microfilaments [176, 181]. Also, cytochalasins, which are well known to disrupt microfilaments, have also been reported to lead to fibronectin release from cells [180, 192].

Immunofluorescence studies implicated an association of extracellular fibronectin fibrils and intracellular actin bundles [183]. It is a very arduous procedure to evaluate and quantitate the concurrence of antifibronectin and antiactin antibodies. This study requires well-spread, low-density cells [183, 184]. The appearance of low-density, well-spread, noncontacted cells may not necessarily be representative of microfilament organization and fibronectin distribution in confluent cultures. Early and some recent electron-microscopic studies have in some cases failed to show an association of extracellular collagen fibrils and intracellular microfilaments in 3T6 cells or collagen-fibronectin fibrils formed by cells grown in ascorbate [112–114, 122]. Other studies showed what was proposed to be an apparent contiguity of intracellular microfilament and extracellular filaments [185]. These studies by Perdue [185] on normal chick embryo fibroblasts appeared to have been somewhat misleading. We now know that the intracellular actin-containing microfilaments are much different from the extracellular filaments seen by electron microscopy which are composed of fibronectin and collagen (or procollagen) and likely proteoglycan [109, 111–114, 119]. Studies from our laboratory [112–114] and those of Singer's [119] using a goniometer stage showed a direct opposition of intracellular microfilaments with extracellular filaments which reacted with fibronectin antibodies. As had been shown in immunofluorescence studies, the extracellular fibronectin filaments were most prominent where intracellular microfilaments were most pronounced—ie, cell-cell and cell-substrate adhesion sites. The studies of Singer were particularly informative using tilted samples to study these intra- and extracellular filaments, which were termed a fibronexus. It should be noted that this extracellular material is not only fibronectin, but also collagen and proteoglycans [115–117, 186].

X. CELL MOVEMENT

The studies cited above deal with extracellular fibrils of fibronectin and collagen and the interrelationship with cytoskeletal structures. Fibronectin will modify cell movement as was initially shown for random migration of fibroblastic cells in vitro [187]. Subsequent studies by Postlethwaite et al [188, 189] showed that fibroblasts were capable of directed motility or chemotaxis.

Additional studies have examined cell movement using a system in which a substrate is coated with colloidal gold particles and cells move about and phagocytose these particles. This assay has been termed phagokinesis, and movement is quantitated by measuring the length of the tracks that are cleared of colloidal gold as the cells move across the substrate [190]. Phagokinetic studies show that fibronectin will promote the random movement of cells [187]. The addition of fibronectin, which presumably winds up largely on the substrate, leads to the movement of cell processes or outgrowth of neurites from the retina [191]. Fibronectin will also promote cytokinesis in serum-free medium [192]. It appears that fibronectin may serve as an anchoring site for cell movement whereby a cell can exert tension at an adhesive site. One apparent paradox regarding fibronectin and cell movement is the observation that transformed cells have increased movement compared to normal cells [reviewed in, 193] while generally having an absence of fibronectin on their surface (see below). It would seem that other adhesion factors must be active in transformed cells and that, since these cells are less adhesive than their normal counterparts, simply increased adhesion does not necessarily correlate directly with increased cell movement. Paradoxically, cell adhesion may be a necessary initial step in the gradient-directed motility or chemotaxis of cells; that is, cells may need to be attached to a given substratum before being able to respond to a chemotactic stimulus. How signals are tranduced from the cells' environment to the cytoskeleton is not clear, but an operational cytoskeleton is necessary for the cells to move about. Drugs that modify calcium and energy production by cells will clearly effect cell movement [see review, 168]. The possible role fibronectin may play in cell movements during development or in wound healing seems evident and is discussed below. For example, studies in lower species and in mammals have shown that in a wound, epithelial cells will migrate out on a fibrin/fibronectin clot formed at the site [194, 195]. Also, fibronectin is present embryologically at sites where cells have moved or where cells will subsequently move [196, 197].

XI. CHEMOTAXIS

In addition to apparently random migration of cells, directional movement of cells is clearly an extremely important function in embryologic development and in the adult animal in dealing with injury, infections and reparitive processes. Because of the known role of fibronectin in coagulation and as an opsonin, it would be of interest to question what role fibronectin may play as a chemoattractant to cells involved in the reparative process.

Following injury, fibronectin will participate in the blood clot (see below) owing to a specific interaction with fibrin and cross-linking via coagulation factor XIII_a [97, 198]. As mentioned above, epithelial cells will migrate over the fibronectin and fibrin in a clot [194, 195]. Clearly, fibronectin itself or chemotactic peptides released from fibronectin would be most useful in bringing inflammatory cells, fibroblasts, smooth muscle cells, and endothelial cells to the area to form granulation tissue and ultimately fibrosis and scar formation. The attraction of fibroblasts to fibrotic regions may be important in fibrosis in scirrhous carcinomas such as is seen in breast cancer. This process may also be important for attracting endothelial and smooth muscle cells during angiogenesis that occurs with tumor formation.

Probably the best-studied cells for chemotaxis are those participating in the inflammatory response, namely neutrophils and monocytes. Many studies have focused on the role of complement as a chemotactic factor, and this will not be discussed here. Of possible significance for many biological problems is the potential role of fibronectin as a chemoattractant for reticuloendothelial cells, monocytes, and macrophages. It currently appears that intact fibronectin is a poor to negligible chemoattractant for monocytes (unpublished observation). Further studies will be addressing the possible role that cleavage products of fibronectin may play in chemotaxis and this should prove informative.

Fibroblasts have also been shown to be responsive to chemotactic stimuli. Studies by Postlethwaite et al [188, 189] reported that fibroblasts respond to chemotactic stimuli produced by lymphokines, collagen, and collagen peptides. Studies by Gauss-Muller et al [199] have attempted to delineate to what extent the promotion of cell adhesion alone or chemotaxis alone plays a role in the overall process of gradient-mediated, directed migration of fibroblasts. An important factor in studies reported on fibroblast chemotaxis is the need for using gelatin-coated filters. When this is done, fibronectin has a major effect on fibroblast chemotaxis [9]. One interpretation of these data is that adhesion to the filters is necessary to promote chemotaxis. Earlier studies showed that fibronectin had an enhancing effect on random mobility of fibroblasts [187]. Because of this observation, it was important to establish to what extent fibronectin stimulated either random migration (chemokinesis) or directed or gradient-mediated migration (chemotaxis) or both. To answer the question, a "checkerboard" analysis was performed to analyze the chemotactic or chemokinetic activity in the filter chamber assay [200, 386]. This particular type of experiment asks the question whether various concentrations of a putative attractant applied above and below a filter (that is, in the presence of a gradient or not) will lead to cells moving through the filter. If cells were to move through the filter when concentrations of attractant are equal above and below the filter, this would not constitute a true chemotaxis, but rather a chemokinesis. It would appear from recent studies that fibronectin can stimulate chemotaxis by 1) increasing adhesion of cells to the gelatin coated membrane, and 2) an apparent stimulation of cell migration independent from adhesion. Other studies will be needed to ascertain orientational direction of fibroblasts to fibronectin or other chemotactic stimuli [200] to distinguish this from simple adhesion. Fibronectin may play a role in haptotaxis, which is movement of cells based on an increasing gradient of adhesiveness.

Fibronectin will serve as a chemoattractant to other cell types, including neural crest cells [210], smooth muscle cells [202], and Schwann cells (Evercooren et al, unpublished observation). Given that Schwann cells or Schwannoma cells produce fibronectin and laminin and form a basement membrane [203] (Palm and Furcht, in press), it is possible to conclude that basement membrane formation and possible chemotactic activity of basement membrane components plays an important role in regenerating neurons following an injury. However, chemotaxis may not only play a role in remyelinating a neurite, as fibronectin can stimulate neurite outgrowth by itself apparently not involving chemotaxis [191]. The mechanism for this phenomenon is not clear at present, but is under investigation. Preliminary studies in our laboratory indicate that laminin and fibronectin will promote migration of Schwann cell derived tumors [McCarthy et al, unpublished]. It also may be possible that both neurite outgrowth and Schwann cell migration may concomitantly ensue following fibronectin or laminin being laid down during developmental changes in the embryo or following injury. Movements of cells to a solid phase fibronectin or laminin would constitute haptotaxis.

In other systems, fibronectin will enhance the killing of tumor cells in vitro by monocytes and macrophages [204]. Clearly, the shedding of fibronectin or peptides from it developed in or around tumor cells may act as chemoattractants to moncytes, macrophages, or possibly neutrophils, which may then have some cytotoxic activity to tumor cells.

On the opposite side of the tumor-host relationship, tumor cells may show directed migration or chemotaxis. Studies from Ward's laboratory have shown that there is a chemotactic factor to tumor cells related to a complement component similar to C5A [205–209]. Clearly, the chemotaxis of tumor cells may contribute to the tissue-specific homing of tumor cells to certain organs. Specifically relative to metastasis, Orr et al [209] have performed studies looking at the chemotaxis of highly metastatic and poorly metastatic strains of fibrosarcoma cells. These studies showed that highly metastatic strains would respond to chemotactic stimuli derived from tumor cells, whereas the poorly metastatic fibrosarcoma cells and normal fibroblasts did not. These data suggested that metastasis and invasion in this one model system may be related to chemotactic responsiveness to a $C5_a$-related peptide that the tumor cell may even participate in generating [209]. It should be noted, however, that the tumors used in this study are likely polyclonal in origin, and may be phenotypically and genotypically quite disparate from one another. It seems that the possible role of chemotaxis in tumor metastasis by hematogenous routes may be fundamentally related to the adhesion of these cells to basement membrane constituents, laminin, and fibronectin, and that chemotaxis might be superimposed upon this initial adhesive event (see below).

The mechanism for generating the chemotactic signal in leukocytes is related to methylation reactions, as inhibitors of this process block chemotaxis [210–212]. Other studies have shown that inhibitors of methylation reactions and the turnover of phospholipids perturb fibroblast adhesion and chemotaxis [199]. Given this, it is reasonable to hypothesize that mechanisms for processing chemotactic signals in granulocytes might be similar to those of fibroblasts. Many of the methylation inhibitors used in fibroblast studies perturbed adhesion of cells, and it is uncertain whether adhesion among collagen, fibronectin, and subsequently cells, necessarily depends on methylation reactions; this awaits further clarification. In addition, it is possible that methylation inhibitors may have a number of other effects on cells unrelated to the effects on methylation reactions.

XII. DIFFERENTIATION AND DEVELOPMENT

Fibronectin has important effects on differentiation and development, and undergoes significant changes during embryologic development. A great deal of work has been performed on the role of fibronectin in myoblast differentiation. An early study by Hynes et al [53] reported that fibronectin (then referred to as LETS protein) increased after myoblast fusion or differentiation. Myoblasts grown in vitro, such as the lines developed by Yaffe, go through stages where cells are "undifferentiated" at low density and then grow to confluence. At confluence or high cell density, myoblasts fuse to form multinucleated myotubes and develop various phenotypic traits of differentiated muscle [214]. Studies from our laboratory showed that a concanavalin A–binding glycoprotein, extracellular matrix, developed on myoblasts as they grow from low density to confluence [118]. When the L6 myoblasts differentiate, the myoblast matrix that bound Con A appeared to be lost or degraded [118]. Subsequently, immunoperoxidase and immunoelectron microscopy studies showed that a fibronectin matrix developed as myoblast cells grew from low to high cell density [56]. As differentiation ensued, which occurs with myoblast cell fusion, the extracellular fibronectin-containing matrix disappeared, and a small amount of nonfibrillar fibronectin was observed on the surface of differentiated myotubes [56]. Immunofluorescence studies showed the disappearance of fibrillar fibronectin associated with myoblast differentiation [57].

Other studies using lactoperoxidase iodination had proposed that myoblasts produced more fibronectin as cell fusion and differentiation ensued [53]. Additional studies have been performed by Podleski et al [215] to examine the extracellular matrix and myoblast differentiation. These studies showed that the addition of cell-extracted fibronectin inhibited myoblast differentiation and cells continued replicating and did not differentiate [215]. Furthermore, proteolytic removal of fibronectin at the appropriate time, presumably at high cell density, led to enhanced cell fusion and differentiation. Also, addition of antifibronectin antibodies enhanced myoblast fusion to form myotubes at this stage [215]. It may be important to note that the studies reported by Podleski and colleagues utilized fibronectin derived from cells, which is possibly fibrillar or insoluble when added back to cells at neutral pH. If soluble plasma fibronectin had this same effect on differentiation, then the fibronectin present in the fetal calf serum in the medium, if sufficient in quantity, might be expected to produce the same effect. All these studies point to the role of the cell matrix in differentiation. Other studies have attempted to modulate differentiation with various drugs or hormones. Initial studies showed that high pharmacologic or low physiological levels of glucocorticoid hormones would block myoblast differentiation and lead to enhanced accumulation of extracellular matrix fibronectin [118, 216]. These studies concluded some concordant relationship between the cell matrix and differentiation in myoblasts. In summary, these studies showed that there is an initial presence of a fibronectin containing extracellular matrix as cells contact one another and that it subsequently disappears as myoblasts fuse together and differentiate.

In related work Puri et al [217] addressed the question of what effect fibronectin per se has on cell fusion as myoblasts differentiate. These studies utilized myoblasts grown in suspension culture, in serum-free medium, and showed that exogenous serum fibronectin had no effect on myoblast fusion in this system. Furthermore, addition of antibodies to fibronectin failed to modulate cell fusion, but did block adhesion of cells to gelatin [217]. Based on other studies cited above, this is not terribly surprising since it was reported that it was the degradation or dissolution of matrix fibronectin at the time of fusion that appeared to be important. Furthermore, the addition of cellular fibronectin blocked fusion and differentiation [215].

Studies concerning these changes in fibronectin in myoblasts may be related to the presence of gangliosides at various stages of myoblast differentiation [218]. As discussed above, studies by Kleinman et al [156] reported that the gangliosides GD_{1a} and GT_1, or related glycoconjugates, may serve as "receptors" for fibronectin. The studies of Whatley et al [218] reported that GD_{1a} increases threefold immediately prior to fusion of myoblasts and returns to baseline low levels in differentiated myotubes. The possibility exists that there is a parallel change in surface glycoconjugates, possibly gangliosides, and matrix fibronectin. It is important to note that the extra-cellular matrix fibronectin, which is by far the largest component of fibro-nectin, would not need to be bound to the ganglioside in the membrane, as would be expected for nonfibrillar or membrane bound fibronectin. Other constituents have been isolated that function as cellular "lectins," and these could potentially modulate myoblast differentiation.

The role of fibronectin in chondrocyte differentiation and the development of cartilage has also been an area of intensive investigation. Studies by Hassell et al [58] and Pennypacker et al [219] showed that differentation of mesenchymal cells into chondrocytes was associated with the loss of surface-associated fibronectin. These studies further proposed that the loss of fibronectin may be necessary for persistence of the chrondroblast-chondrocyte phenotype. Other studies reported that fibronectin was absent from differentiated cartilage and chondroblasts in sternal cartilage [220]. When cartilage cells are dissociated and plated in vitro, the chondrocytes appear to synthesize fibronectin. As chondrocytes differentiate and synthesize an extracellular cartilage matrix, fibronectin then disappears [220]. More recent studies by Dessau et al [221] examined type II collagen, fibronectin, and chondroitin sulfate proteoglycan synthesis and secretion during chondrocyte differentiation. These and other studies have shown the disappearance of fibronectin as the cartilage matrix is formed [219–221]. Other studies examined the synthesis and distribution of fibronectin during the differentiation of cartilage and bone in rats [222]. These studies utilized demineralized bone injected subcutaneously into newborn rats which led to the development of cartilage, bone, and bone marrow. This study reported maximal synthesis of fibronectin in the early stages of the collagenous matrix-mesenchymal cell interaction. Furthermore, fibronectin was reported to be continually synthesized during chondrogenesis, osteogenesis, and hematopoiesis of the injected matrix material [222]. Studies by West et al [223] reported that fibronectin reversibly blocks chondrocyte differentiation, and concomitantly induces a fibroblastic phenotype in chondroblasts or mesenchymal cells. This is also consistent with studies by Linder et al [41] which showed the disappearance of fibronectin from chondrogenic regions of chick limb buds and areas of embryonic cartilage. This is similar to the effect of glucocorticoid hormones which block myoblast differentiation and lead to fibronectin accumulation [118, 216]. In related studies it was shown that retinoids would block chondrogenisis and enhance the apparent amount of surface-associated fibronectin [58].

Clearly, fibronectin can serve a function related to adhesiveness and phenotypic traits of various mesenchymal cells including chondrocytes. It is also of note that another protein has been isolated from plasma that will promote the attachment of chondrocytes but not fibroblasts to collagen [133]. This protein chondronectin is distinct from fibronectin; it is a glycoprotein with a molecular weight of 180,000 daltons, and migrates as disulfide-bonded subunits [133]. Chondronectin may have important effects on differentiation, and may act in an opposite manner to fibronectin in its effect on mesenchymal or chrondrocyte cell phenotype; this awaits further investigation.

In lower animals such as the sea urchin embryo fibronectin or a molecule antigenically related to fibronectin was found on the cell surfaces and between cells at blastula and gastula stages [224]. In the studies by Spiegel et al [224] during gastrulation in the sea urchin embryo, there is a decrease in fibronectin on the outer cell surfaces. At this stage, the cells of the ectodermal layer elongate as they migrate, and this may contribute to the apparent decreased fibronectin observed. However, at the same time in sea urchin embryos, fibronectin intensity appears increased at the invagination site. In other studies, a surface constituent appeared during reaggregation experiments on fresh-water sponge cells which cross-reacted with antibodies to human fibronectin [225]. This raised the possibility that a fibronectin like molecule may be present on the surface of cells of very primitive organisms and be responsible for adhesive interactions and morphogenisis. This molecule from sea urchins may not be identical with vertebrate fibronectin, but clearly it must have regions or epitopes that are recognized by antibodies that have some sequence homology with mammalian fibronectin.

In higher organisms such as the chick, fibronectin was observed in the ectoderm of very early chick embryos and was proposed to modulate morphogenetic movements occurring with differentiation and development [196]. Other, earlier studies showed that fibronectin, then called "fibroblast surface antigen" [41], changed or was lost from primitive mesenchymal tissues as they differentiated. In the embryonic chick eye, fibronectin is largely found in basement membranes such as the Decemet membrane in the cornea, and is also present in the sclera, vitreous body, and primitive mesenchyme [226]. It appears that fibronectin is of mesenchymal origin in the eye, as it was not detected in primary epithelial stroma but was in the secondary stroma which had been invaginated by mesenchymal cells. Additionally, fibronectin was found associated with migrating corneal endothelial cells. As the cornea matured, fibronectin was observed to disappear, although in the mature cornea fibronectin remains in the Descemet membrane [226].

Recent studies by Akers et al [191] showed that fibronectin produced a significant promotion of neurite outgrowth from embryonic chick retinal cells. In other studies, immunoperoxidose experiments indicated that fibronectin was present in the basement membrane surrounding neural crest cells [227]. Chick neural crest cells were observed to migrate out upon a fibronectin-containing matrix. Some of the neural crest cells were interpreted to have synthesized fibronectin, and those that migrated were thought to be those that first synthesized fibronectin [227]. These fibronectin-positive neural crest cells may have been of mesenchymal cell origin, however. Other studies reported that neural crest cells cultured in vitro required fibronectin to attach to various types of collagen (I–V) [228]. Also, some very elegant studies showed that fibronectin stimulated chemotaxis and chemokinesis of neural crest cells in Boyden chambers [228]. Other in vitro studies showed that addition of fibronectin to quail neural crest cells promoted differentiation to adrenergic neurons [229]. Fibronectin and laminin have recently been shown to differentially promote chick peripheral central nervous system neuron outgrowth (Rogers et al, unpublished).

Studies by Waterman and Balian [230] examined fibronectin in the oral cavity of the developing chicken. It was observed that fibronectin is in the floor of the foregut at the stages of formation and rupture of the oral membrane. Additionally, fibronectin was observed 1) in dorsal mesocardium, 2) in the Rathke pouch and infundibulum, and 3) within the closing plates between ectodermal clefts and endodermal pouches.

Fibronectin appears to be involved in differentiation of tooth germ cells as shown in experiments using mice. Dental mesenchyme loses fibronectin as cells differentiate into odontoblasts [231]. Fibronectin was not present in the mineralized dentin or epithelial tissues. Fibronectin was concentrated at the interzone between dental epithelium and mesenchyme, where it was proposed to be involved in the adhesion of mesenchymal cells as differentiation to odontoblasts occurs [231].

A number of studies in mouse embryos have reported that fibronectin changes with development and differentiation [232, 233]. These studies in early mouse embryos reported that fibronectin was not present on preimplantation, two-, four-, or eight-cell embryos, on morulas, or the trophectoderm of blastocysts. Fibronectin was not observed on the inner cell mass isolated from early blastocysts [232]. However, 24 hours later, when the inner cell mass differentiates into endoderm and ectoderm, fibronectin then appears. Therefore, the appearance of fibronectin in early mouse embryos appears as an early differentiation marker. In contrast, similar studies by Wartiovaara et al [233] failed to find fibronectin in the preimplantation mouse embryo, but did observe fibronectin on the inner side of the trophectoderm of the implanting $4^{1}/_{2}$-day embryo. The conclusions of the studies by Wartiovaara et al [233] were that 1) fibronectin was expressed in the late blastocyst coincident with the appearance of endoderm, 2) fibronectin accumulated between different germ layers, and 3) expression of fibronectin was a mesodermal product. Additional studies on murine teratocarcinoma cells showed that differentiation of embryonal carcinoma cells to endoderm is associated with the expression of fibronectin [232].

XIII. TRANSFORMATION

A major driving force in the study of fibronectin, mentioned briefly in the introduction, was and is the well-documented decrease of fibronectin from the surface of many transformed cells. In the early 1970s, a number of laboratories, roughly simultaneously, determined that there was a transformation associated decrease in a glycoprotein that we now have come to refer to as fibronectin. The first report of a transformation-associated decrease in surface fibronectin was seen in the studies of Hynes who subsequently referred to this protein by the acronym LETS, large external transformation sensitive [234, 235]. Other studies dealing with the same problem referred to this protein as galactoprotein A [236, 237], Z protein [238], LI protein [239], cell-surface protein (CSP) [17], and cell-surface antigen SFA [240–242].

Two studies, among others, provided important tools for progress in this area of fibronectin and noncollagen glycoprotein research in general. The first, by Ruoslahti and Vaheri [241, 242], using immunodiffusion experiments, showed the identity of a high molecular weight chick fibroblast surface protein with a serum protein which migrated as an alpha-globulin, and is now called fibronectin. The presence of fibronectin in human plasma at high levels of approximately 300 µg/ml then provided a readily accessible source for isolation and purification. This was followed by the method of gelatin-sepharose affinity chromatography of Engvall and Ruoslahti [29], which permitted the relatively easy, rapid isolation of large quantities of plasma fibronectin that could be submitted to extensive biochemical analyses. Another major advance technologically was the pioneering study of Yamada and Weston [17] that isolated cell matrix fibronectin from cell layers of normal chick fibroblasts by urea extraction. These studies then provided a tool to study the soluble and matrix form of fibronectin and possibly ascertain what molecular differences there may be. Additionally, the elucidation of the biochemical difference between soluble and insoluble fibronectins would be applicable to understanding the possibility of a molecular perturbation in fibronectin produced by transformed cells.

Various investigative avenues have been pursued in an attempt to define why fibronectin is absent from the surface of transformed cells. Some of the earliest studies dealt with observations that there were increased levels of proteases in transformed or malignant cells [see reviews in 243]. Shortly thereafter, Hynes [235] reported that the decrease in fibronectin seen on transformed cells could be mimicked by protease treatment of normal cells. Subsequent studies showed that if matrix fibronectin was added to transformed cells it would cause a phenotypic reversion toward normal which was manifested as increased spreading and adhesion to the substrate [136]. A coordinate set of changes occurs when cell derived fibronectin is added to transformed cells including a flattened more normal morphology [135, 136], increased adhesiveness [136], a more organized cytoskeleton [137], and decreased surface modifications such as microvilli or membrane ruffles [244]. Importantly, however, the addition of exogenous cellular fibronectin to transformed cells did not restore contact inhibition of growth to these cells.

Fibronectin is decreased in cells transformed by various viruses, both RNA and DNA, temperature-sensitive mutants of oncogenic viruses, and chemically transformed cells when compared to seemingly normal counterparts (see Table I).

Various normally occurring phenomena or exogenous treatments can modulate surface fibronectin so that normal cells appear more like transformed cells or that transformed cells more resemble normal cells. For example, as normal cells go into mitosis, the cells transiently become less adhesive to the substrate, lose the matrix associated fibronectin, and round up and divide [234, 245, 246]. Also, NRK cells which go into mitosis, lack fibronectin and collagens detected by immunofluorescence [110]. The paradox in this cell cycle effect is that cells in serum-free medium require fibronectin for normal cytokinesis [192]. Treatment of normal cells with thrombin increases fibronectin in the medium probably by increasing production and release [162]. Treatment of normal cells with other proteases including trypsin, plasmin, elastase and others appears to cleave matrix and membrane associated fibronectin and also microfilaments [181, 235, 247–250]. There is no clear-cut association of release of fibronectin and mitogenisis; that is, release of fibronectin with certain proteases will be associated with mitogenisis.

TABLE I. Transforming Agents Associated With Changes in Fibronectin

DNA virus
   SV_{40}
   Polyoma virus
   Adenovirus
   Others
RNA virus
   Avian (Rous) sarcoma virus
   Avian leukemia virus
   Avian leukosis virus
   Murine sarcoma virus
   Hamster sarcoma virus
   Others
Chemical

However, treatment of normal growth arrested cells with thermolysin or papain will remove matrix fibronectin but does not cause cells to proliferate. Though fibronectin is exquisitely sensitive to a variety of proteases, the supenatant fibronectin from cultures of transformed cells appears to be intact as determined by reduced SDS polyacrylamide gels (see review Vaheri and Mosher [2] and references therein).

A number of other agents can cause the release of fibronectin from the surface of normal cells. For example, the tumor promoter phorbol myristate acetate (TPA or PMA) has been reported to cause the gradual loss of fibronectin from cell layers [251, 252]. In preliminary studies we have not been able to confirm the release of fibrillar fibronectin using normal human fibroblasts treated with the tumor promoter PMA (Furcht, unpublished observation). The mycotoxin, cytochalasin B has well-known effects on cell shape, cytoskeleton, and glucose transport. In many ways, cells treated with cytochalasin B resemble protease treated normal cells or mitotic normal cells relative to shape, adhesion, and cytoskeleton, most specifically microfilaments. As discussed above, cytochalasin B treatment of cells has independently been reported to release surface fibronectin from cells. Upon removal of the cytochalasin or following recovery from protease treatment, cells will redevelop an extracellular fibronectin matrix. Other work has shown that disulfide reducing agents decrease the amount of fibronectin in radiolabeling experiments, though immunofluorescence studies after cell treatment with disulfide reducing agents showed only a negligible decrease in matrix fibronectin [253].

It has also been reported that a 10,000-dalton polypeptide derived from the conditioned medium supernatants of fibrosarcoma cells is capable of releasing fibronectin from matrices without degrading the fibronectin [254]. This is an intriguing phenomenon, as this polypeptide has protease activity and the effects of increasing release of fibronectin by the 10,000-dalton polypeptide are similar to those observed following thrombin treatment of normal human cells [255]. It will be important to further characterize this or related cell surface active agents which may be produced by transformed or malignant cells which can alter the cell matrix. It is uncertain at present, but this phenomenon may be important in metastasis or in homing of cells.

The synthesis of fibronectin is decreased in transformed avian cells [256], and this is discussed in another section. By contrast, the level of fibronectin synthesis does not appear grossly abnormal when comparing appropriate normal and transformed human cells, though undoubtedly certain transformed human cells may have decreased synthesis [255].

From these and other studies it would therefore appear that at least in certain transformed human cells that synthesize and secrete significant amounts of fibronectin into the medium, that there is an abnormality in processing of fibronectin or of its binding to the cell surface—ie, possibly an abnormality in a "receptor" for fibronectin. Also, if multiple fibronectin genes exist, perhaps a "cell" fibronectin gene may be turned off in transformed cells. Recent studies by Wagner et al [257] have provided additional information on the similarities and differences between fibronectin produced by normal and transformed cells. Normal and transformed cell fibronectin was intact and dimeric; it bound to gelatin, bound to thiol sepharose, bound to cells, promoted cell attachment, and gave similar peptide maps. Importantly, transformed cells were defective in binding exogenous fibronectin derived from normal or transformed cells. This has implications for what the receptor or cell surface binding site for fibronectin may be. It may be related to defective or anomolous production of proteoglycans by transformed cells (see below). Carbohydrate analysis showed more branches per core and increased sialic acid on asparagine-linked sugars of fibronectin from transformed cells [257].

A number of recent reports showed that fibronectin could be restored to the surface of transformed cells. The first report using transformed human fibroblasts showed that treatment with glucocorticoid hormones, such as dexamethasone, led to the development of an extracellular fibronectin and collagen matrix [258, 259]. The glucocorticoid-induced matrix is associated with increased cell adhesion, with an increase in production of fibronectin detectable in the medium, and most importantly a 500% increase in fibronectin in the cell layer detected by ELISA [258–260]. The glucocorticoid-induced phenotypic reversion of these cells is a direct effect, not requiring other hormones or serum factors; it requires RNA and protein synthesis, and occurs most prominently when cells are at high density [260]. Recent preliminary studies performed in collaboration with Drs. N. Oliver and S. Bourgeois have suggested as much as a 10-fold increase in the rate of synthesis of fibronectin in dexamethasone treated transformed cells (unpublished observation). The glucocorticoid effect on increasing matrix-associated fibronectin has also been reported for other cell types, namely hepatocytes [261]. Other drugs or substances will increase fibronectin synthesis or its being laid down in a matrix form on transformed cells include 1) butyrate treatment [262], 2) interferon treatment [263, 264], and 3) cAMP treatment on CHO cells [265].

In a different model, it was reported that growth of 3T3 cells in low serum concentrations led to the loss of fibronectin from the cell surface, and that epidermal growth factor blocked this loss [266]. These 3T3 cells may be peculiar in the response to serum restrictions, as other cells do not seem to respond to serum restriction by a loss of fibronectin [267]. Also, other studies from Yamada's laboratory reported that continuous "normal" cell lines have substantially less fibronectin than truly normal primary or secondary cell cultures [268].

Fibronectin is only one component of the extracellular matrix, which also includes proteoglycans, collagens, and other noncollagenous glycoproteins. Since these components may all interact, an abnormality of one may lead to a corresponding change in the others. It is well known that various transformed cells have decreased collagen production [256], and other studies showed that abnormal types of collagen are synthesized by various transformed cells [269].

Studies by various laboratories have shown the production of proteoglycans by cells in culture [117, 270–274; see review, 151]. An important observation is that normal cells have more heparin sulfate on their surfaces than the transformed counterparts [275]. Other studies have shown that heparan sulfate may have components such that a proportion exists as an intrinsic membrane component [161] (see section on membrane receptor for fibronectin). A plausible working hypothesis is that transformed cells have decreased heparan sulfate or collagen on their surface and therefore do not have the appropriate attachment site for the "nucleation" of other matrix components, and transformed cells therefore fail to form a matrix. As discussed above, complex gangliosides GD_{1a} and GT_1 may serve as binding sites for fibronectin. Some time ago studies from Hakomori's laboratory showed that transformed cells have a defective synthesis of complex glycolipids [236]. It is possible that quantitative or qualitative abnormalities in any one of the proteoglycans, collagen, noncollagen glycoproteins, or glycolipids serves as an initiating factor which then leads to secondary abnormalities in the others resulting in the transformed phenotype. Importantly, there are exceptions to the general observations that fibronectin is lost from the surface of transformed or malignant cells [276–280].

XIV. MOLECULAR BIOLOGY/GENE MAPPING

The large size of the fibronectin subunit chain has hindered advancement in isolation and analysis of the fibronectin gene or genes. At the chromosomal level, a variety of studies have been performed that have attempted to define the locus for fibronectin production. Studies by Eun and Klinger [281], Smith et al [282], and Ruoslahti and Klinger [283] reported that chromosome 11 appeared to be the locus for fibronectin production in human-human and human-nonhuman (eg, mouse) heterokaryons. Other reports by Owerbach et al [284] and recently by Rennard et al [285] found that human chromosome 8 was responsible for fibronectin production. Other studies postulated that human chromosome 11 was the structural locus for fibronectin production, and that chromosome 3 may have a locus that serves as a regulator for fibronectin expression [282]. One of the potential problems in some of the studies is the possibility that the serum in which cells are grown could contribute the fibronectin to heterokaryons which might not otherwise have it. Clearly, species-specific polyclonal and, more directly, species-specific monoclonal antibodies are required to evaluate xenogeneic heterokaryons correctly in these studies. Additionally, no studies to date have addressed the question whether unique genes exist for plasma, interstitial, or basement membrane fibronectins, and, if they do, whether these are on the same or different chromosomes.

Studies by Fagan et al [286] constructed various recombinant plasmids which contained cellular fibronectin cDNA sequences. The cDNA sequences were complementary to the corresponding cellular fibronectin mRNA by hydridization studies, and a sequence determination of the putative gene was subsequently performed. An earlier report found that the translatable mRNA for fibronectin was decreased in Rous sarcoma virus–transformed cells [256]. Subsequent studies by Fagan et al [286] partially purified the RNA for cell fibronectin that had a molecular weight of 13×10^6 daltons, and found that specific fraction was profoundly decreased in Rous sarcoma virus–transformed fibroblasts. Recently, Fagan et al [287] used cellular fibronectin $_c$DNA as a probe and showed that the specific mRNA for fibronectin was only 10–13% of normal levels in transformed chick cells. It will be important to ask the same question regarding transformed human cells as those cells synthesize levels of fibronectin approximating normals but fail to lay it down in a matrix form.

XV. FIBRONECTIN IN COAGULATION AND PLATELETS

Early studies on plasma fractionation pointed to certain unique properties of what we now term fibronectin. Fibronectin will coprecipitate with fibrinogen-fibrin complexes [288], which are seen in cryofibrinogenemia and disseminated intravascular coagulation [12]. Also, fibronectin will participate in the heparin-precipitable fraction in plasma at reduced temperatures [289]. This heparin-precipitable fraction generally includes fibronectin and fibrinogen; however, fibronectin will precipitate alone at reduced temperature with heparin, but fibrinogen will facilitate or augment this precipitation [289].

Fibronectin will bind to affinity columns of fibrin and fibrinogen, though there is greater binding of fibronectin to fibrin when this is carried out at room temperature rather than at 4°C. The level of fibronectin is 20–50% lower in serum than in plasma [13], and is significantly lower in critically ill patients, particularly those with disseminated intravascular coagulation [290]. In vitro, if clotting occurs at 0–4°C there is increased binding of fibronectin to the clot [291], which is due to noncovalent association and covalent transglutamination [292]. The covalent linkage is due to the action of coagulation factor XIII$_a$ forming gamma-glutamyl cross links with lysine residues between fibronectin and fibrin [15, 292]. Fibronectin can be cross-linked to collagen by a similar mechanism [293]. The cross-linking site has been ascribed to the amino terminus of the fibronectin molecule [294]. Also, the carboxy terminal region of the alpha chain of fibrinogen is necessary for this cross-linking to occur [288, 289]. Fibronectin may be cross-linked to itself or other matrix constituents in cell layers [164]. Interestingly, there is a difference in the transglutaminase activity between normal and transformed cells, and this might somehow be related to the absence of matrix fibronectin in transformed cells [165]. Finally, polyamines and dansyl cadaverine will inhibit the cross-linking of fibronectin with various other substrates using factor XIII_a [295].

Platelets contain 2–4 μg of fibronectin per 10^9 platelets [296]. Stimulation of platelets with collagen or thrombin leads to a release of fibronectin presumably from within platelet alpha granules [297, 298]. Associated with aggregation, fibronectin can be visualized on platelet surfaces [79, 296, 298, 299].

Studies by Bensusan et al [73] proposed that fibronectin was the collagen receptor on platelet plasma membranes. There appears to be some plasma membrane associated fibronectin on platelet surfaces from data on immunoelectron microscopy [79]. Corroborating this study, experiments using antifibronectin antibodies showed that these antibodies stimulated platelet secretion [73]. In contrast to this, other studies reported that there was little or no fibronectin on the platelet plasma membrane [298, 299], though it is not entirely clear how sensitive these assays were. Though a number of studies have shown that fibronectin is present on the surface of platelets, this does not necessarily mean that fibronectin serves as the "receptor" for collagen. Other platelet surface constituents may interact with collagen and serve as a binding site for collagens and not necessarily be a well-defined saturable receptor. Additionally, if fibronectin were the only receptor for collagen, it is unclear why denatured collagen, which has a higher affinity for fibronectin than native collagen, does not aggregate platelets [300]. This further confounds a clear-cut explanation regarding the role of fibronectin in platelet aggregation. Studies reported that antibodies to fibronectin would produce a small decrease in the adhesion of platelets to native collagen [75]. Also, preincubation of platelets with gelatin followed by subsequent stimulation with native collagen produced a decrease in aggregation, but not a complete blockade, as might be expected if fibronectin were the sole receptor for native collagen on platelet surfaces. This last observation led Santoro and Cunningham [75] to propose a multivalent interaction theory, which states that there are multiple binding sites for collagen on platelet surfaces, none of which may be of tremendously high specificity but which collectively lead to an apparently specific interaction. In contrast to the data on collagen-binding sites on platelets, there are inducible, saturable receptors for fibrinogen on the surface of platelets following ADP stimulation [301]. It is not certain whether this represents fibronectin or not. Other studies reported that plasma fibronectin does not appear to be involved in ADP-induced aggregation of platelets, although fibronectin would undoubtedly be released during ADP aggregation [297]. One potential explanation for what appears to be conflicting data on the role of fibronectin in platelet function is the possibility that fibronectin 1) is involved in platelet adhesiveness more so than platelet aggregation, and 2) may be only a minor component in the sum of various molecular interactions on the surface of platelets that leads to their adhesiveness during aggregation.

In this regard, platelets have been known for some time to be adhesive to collagen [302–306]. Most studies have utilized type III or type I collagen, which must be in a native state to induce platelet aggregation or adhesion [300, 307]. Also of note, specific peptide fragments of collagen chains appear responsible for the interaction with platelets [308, 309]. It is important to note that studies of Koteliansky et al [310] found that fibronectin would modulate platelet adhesion to collagen.

Various studies have looked at platelet adhesion to basal lamina in vitro in an attempt to evaluate what may be occurring during the initial phases of thrombus formation [311–313]. It is clear that adhesion and aggregation of platelets may be separable phenomena. Furthermore, studies by Huang and Benditt [313] examined platelet adhesion and degranulation using glomerular basal lamina, the noncollagen or collagenous matrix of basal lamina or tendon collagen fibers. These studies reported that platelets adhered predominately to the noncollagenous matrix components of basal lamina, and that adhesion and spreading did not necessarily result in degranulation. Also, basal lamina collagen (presumably type IV and possibly some type V collagen) was not effective in promoting platelet adhesion. Other studies have examined the other genetically distinct forms of collagen and their ability to aggregate platelets [314].

Other reports have implicated the Von Willebrand factor in platelet adhesion [315]. More recent studies reported that platelets adhere to arterial subendothelium using the Von Willebrand factor/factor VIII complex [316]. Fibronectin has a similar subunit molecular weight to that of Von Willebrand factor, which is also a glycoprotein [15]. Patients with Von Willebrand disease have a bleeding disorder, one component of which is a defect in platelet adhesion. Fibronectin, like Von Willebrand factor, is found in platelets and in the subendothelial connective tissue matrix, and, importantly, Von Willebrand factor will bind to collagen [317, 318]. It is likely that both fibronectin and Von Willebrand factor are exposed to platelets at sites where endothelial cells are detached or at sites of vessal damage and injury. The possible interactions of fibronectin and Von Willebrand factor and the degree of adhesion each of these may contribute to platelets await further definition. There are other defects in platelet adhesion such as the Bernard Soulier syndrome or Glanzmann thrombasthenia. These diseases have an abnormality that does not seem to be in fibronectin or Von Willebrand factor, but rather may have some other perturbation in cell surface glycoproteins [319].

Relevant to the role of fibronectin in platelet function, we studied a kindred with Ehlers-Danlos syndrome that had a platelet aggregation defect to collagen [77]. This family appeared to have recessively inherited Ehlers-Danlos syndrome with hypermobile joints, excess skin distensibility, abnormal wound healing and scar formation, and a platelet dysfunction. The affected individual and other members of this family had normal immunoreactive levels of plasma fibronectin. The in vitro platelet aggregation defect to collagen observed in this family could be corrected by adding purified fibronectin from normal plasma [77]. Specifically how seemingly defective plasma or platelet fibro-nectin may be related to connective tissue fibronectin and the relationship to the connective tissue disease in this family is unknown. It remains to be proved what the precise molecular defect may be in the fibronectin of these particular Ehlers-Danlos individuals, but a perturbation in the interaction of fibronectin with heparin, proteoglycans, or collagen could be very important in the platelet dysfunction, abnormal wound healing, and connective tissue abnormality seen in this and similar families. Preliminary studies indicate a change in the migration patterns of proteolytic fragments of fibronectin from these patients on two-dimensional gels (unpublished observation).

XVI. OPSONIC FUNCTION

The reticuloendothelial system functions to clear the body of damaged tissue, old cells, foreign organisms, and cells. The immune system involving antibody, complement, and cell-mediated functions provides highly specific methods to deal with microorganisms and abnormal cells. Prior to the development of a specific immune response to microorganisms, there are clearly nonimmune, humoral, and cellular events involving the reticuloendothelial system which can deal with the various insults to the body. The reticulo-endothelial system includes, among other cells: monocytes in blood; macrophages, the largest component of which is in the liver called Kupffer cells; macrophages in other areas including the spleen, bone marrow, lymphatic system, and possibly phagocytic cells in the central nervous system [32]. The reticuloendothelial system is important in removing cellular debris following trauma, surgery, or wounds that produce damaged tissue, fibrin, and other coagulation components, collagen, dead cells, possibly bacteria, and various other particulate material or altered proteins [321].

Colloidal particles such as gelatin, carbon, and a gelatin-coated colloid coupled to an isotope or some other marker, are readily taken up by the reticuloendothelial system. Additionally, it was observed that plasma factors were somehow involved in this process [reviewed in 322, 323]. For example, this was shown by injecting an animal with gelatin colloid and observing that when another injection was given within a relatively brief time period, there was little uptake of the new colloid. This phenomenon was termed a "blockade" of function of the reticuloendothelial system [321, 324]. It was observed that there was a plasma factor which was depleted or consumed following the first injection of colloid [320–322]. This plasma factor was termed α_2-opsonic protein by Saba and colleagues who began working extensively on this phenomenon in the late 1960s and early 1970s. This protein was thought to be nonspecific—that is, not antibody-dependent. A better term might be multispecific opsonin, or nonimmune opsonin.

The initial in vitro studies quantitated levels of this opsonic factor using a bioassay that quantitated the uptake of a lipid emulsion by liver slices. Prior studies had shown a plasma factor responsible for carbon uptake by perfused liver [325]. It was shown that levels of this factor decreased in plasma following injection of a colloid, and that this factor could be removed by incubating plasma with gelatin colloid [322]. Furthermore, the plasma factor could agglutinate gelatin coated particles or erythrocytes [326]. It was shown further that the opsonic activity exhibited by this protein was contingent upon the presence of heparin or some component present in heparin preparations [34, 327]. As discussed above, fibronectin has been purified by affinity chromatography on gelatin-Sepharose [29]. Importantly, care must be taken to not denature the protein with elution buffers as it may no longer be opsonically active after harsh treatments and denaturation which are sometimes used [34].

Clearly, fibronectin plays a significant role in macrophage and reticuloendothelial cell function. It has been shown that macrophages will synthesize their own fibronectin and are capable of binding exogenous fibronectin [80, 81]. It has also recently been shown that monocytes and macrophages have receptors for fibronectin [82]. It has been postulated, largely by Saba and colleagues, that depression of plasma fibronectin levels could lead to decreased reticuloendothelial system function and predispose to organ failure, sepsis, or other complications which would compromise the patient or animal [322].

The role of fibronectin in coagulation and opsonic function necessitated the development of assays which could rapidly measure levels in plasma. Levels of fibronectin have been evaluated by the functional bioassay using liver slices or, more recently, with macrophages or macrophage cell lines. Immunochemical methods using enzyme-linked immunoassays, Laurell rocket immunoassays, radioimmunoassays, and kinetic nephelometry can be performed. Each method has its own particular utility and specific application. Enzyme-linked immunoassays are best suited to studies where low amounts of antigen are present, and/or when many samples may be involved, such as studying various conditions of cells in culture. Laurell rocket immunoassays were among the first immunoassays for fibronectin, and are readily performed on animal and human plasma samples [328]. Laurell rockets or electroimmunoassays, as they are sometimes called, are quite sensitive and do not require tremendous equipment costs. However, this assay takes approximately a day to perform, requires significant amounts of antibody, and is sometimes difficult to adapt to do very large numbers of samples. Nephelometric assays are based on light scattering due to antigen antibody complexes and are significantly more sensitive than turbidometeric assays. In kinetic nephelometry, the rate of antigen antibody complexes is measured and compared to standards and allows for extremely rapid evaluations. Kinetic nephelometry assays may be performed in a very brief amount of time, such as in 5–10 minutes. A rapid assay is useful in some critically ill patients so that determinations can be made as quickly as possible. The possible disadvantages of the nephelometric assays are that 1) it takes significant amounts of high titer antisera; 2) fibronectin specifically must be measured in plasma, not serum; and 3) and plasma can cause problems with polyethylene glycol, which is used in the reaction mixtures for nephelometric assays (unpublished observation).

Most of the data on plasma levels have come from bioassay and Laurell rockets. This has shown that fibronectin (α_2 opsonic protein, cold-insoluble globulin) levels fall in animals or humans that have had major trauma or injury, severe burns, following major surgery, in certain cases of sepsis or cancer, and in various other critically ill patients [290, 322]. The normal levels are approximately 300 μg/ml, with the levels being slightly higher in men than women [290, 329]. The levels in plasma are higher than in serum owing to the cross-linking of fibronectin to itself and fibrin by factor XIII_a, binding of fibronectin to fibrin and possibly some proteolytic cleavage by coagulation factors—eg, thrombin and plasmin. A common theme in cases of hypofibronectinemia is the occurrence of disseminated intravascular coagulation or significant tissue destruction presumably stimulating the binding of fibronectin to fibrin, collagen, and damaged cells, thereby leading to its consumption [290, 322].

It would seem that some conformational change or oligomer formation of the molecule must occur subsequent to binding damaged tissue, which then permits the molecular complex to be opsonized by the reticuloendothelial system. Specifically why fibronectin would bind to abnormal (damaged) cells or effete tissue as opposed to normal is not known. Why this leads to a recognition phenomenon by the reticuloendothelial system may be due to quantitative considerations and possibly an alteration in fibronectin subsequent to its binding cells which then leads to phagocytosis by reticuloendothelial cells.

The decreased levels of fibronectin complicating various diseases have been proposed to predispose to organ failure, and this may be a physiologic sequela which may occur in the conditions described above. Since fibronectin is present at high levels in cryoprecipitate, Saba and colleagues embarked a number of years ago on a series of studies to use cryoprecipitate therapy in animals and humans to increase fibronectin levels and to ascertain whether there was a commensurate improvement in reticuloendothelial function, patient status, and other physiological phenomena. Cryoprecipitate is reported to correct the apparent opsonic deficiency seen in septic patients with burns or trauma or following major surgery. Using patients with sepsis and organ failure following severe trauma, Scovill, Saba, and colleagues [330, 331] performed a variety of studies to evaluate the physiological parameters before and after cryoprecipitate therapy. It was observed that ventilation of pulmonary dead space (areas that were not well perfused with blood) was high at time zero, and following administration of cryoprecipitate this decreased in the patient group as a whole. In concert with the improvement in ventilation of dead space the arteriovenous shunt decreased substantially from 4 to 48 hours after cryoprecipitate therapy. Further studies showed that limb blood flow, which was used as a parameter for perfusion, increased following cryoprecipitate administration, and that oxygen delivery to tissues increased [331].

The reason for the decreased levels of fibronectin in various clinical conditions may be related to alterations in the turnover, consumption, distribution, and synthesis of plasma fibronectin. Clearly, fibronectin will bind to injured cells and collagen, and be laid down with fibrin during coagulation. These processes would all lower levels of fibronectin due to consumption. There could be decreased levels of synthesis of fibronectin which might be related to sepsis or perhaps poor perfusion of organs. Studies on turnover rates have not been unequivocally defined, though based on the cryoprecipitate infusion experiments one would suspect the $T^{1}/_{2}$ to be greater than 24 hours [332, 333]. It would seem at present that the most plausible explanation for the decreased levels in many clinical conditions is consumption or redistribution; that is, fibronectin binding to damaged tissue, fibrin, etc.

During these processes, many degradative enzymes are released from inflammatory cells, neutrophils or macrophages, along with activation of plasma proenzymes to yield active enzymes such as plasmin and thrombin. This may contribute to increased consumption along with generating proteolytic fragments of fibronectin, which might 1) act to block an activity of the intact molecule, or 2) have a novel function not present in the intact molecule. Relative to the former, recent studies showed plasmin fragments of fibronectin could produce a blockade of reticuloendothelial function [334]. Relative to the latter case, a fragment of fibronectin has been isolated that will function as an opsonin whereas in this particular system the intact molecule does not [336].

A number of comments should be raised in the studies on cryoprecipitate administration. First, it has not been clearly shown that the levels of fibronectin that occur in severe burns, trauma, or major surgery leads to an in vivo abnormality in the reticuloendothelial system in humans. Second, cryoprecipitate contains numerous constituents other than fibronectin such as fibrinogen and coagulation factor VIII. The levels of fibronectin in seriously ill patients are rarely under 100 μg/ml [290], and it has not been determined that this level of fibronectin would be rate-limiting in opsonic activity of the reticuloendothelial system. The in vitro assays that test opsonic activity using liver slices or monocytes show that levels significantly below 100 μg/ml final concentration can still lead to opsonization. The thesis for the utilization of cryoprecipitate in certain critically ill patients presumably harks back to the experiments of injecting colloid, and getting reticuloendothelial uptake; then, when challenged with a second infusion of gelatin colloid, a reticuloendothelial blockade occurs. It would seem that the "blockade" would be a function of the amount of gelatin infused initially, the level of plasma fibronectin, other factors effecting the level of plasma fibronectin such as proteolysis, etc, and the amount of gelatin in the second or challenging dose.

One would suspect that animal experiments could be devised to produce decreased fibronectin, perhaps without producing a reticuloendothelial "blockade" or "saturation." This would permit an evaluation of what other physiological alterations must be going on in patients or animals at the time of the decreased fibronectin—eg, sepsis, shock, etc—that contribute to a relative opsonic deficiency, if it occurs in vivo. In this regard, we have recently observed that childhood acute lymphoblastic leukemia patients treated with L-asparaginase have the lowest levels of fibronectin we have observed in humans, and lower plasma levels than those reported routinely in various other clinical conditions with extremely low levels [290, 322] (Furcht et al, unpublished). In L-asparaginase-treated patients, we have seen fibronectin levels of 50–75 μg/ml, which in some particular patients represented almost a 90% reduction. Though it was not specifically tested in vitro, this was not associated with any outward sign of reticuloendothelial dysfunction, and the patients have no undue problems with infections. These patients have low levels of many plasma proteins including, among others, fibrinogen and complement and manifest problems with a coagulopathy which may be related to decreased activity of coagulation inhibitors [335]. The sequelae of reticuloendothelial dysfunction in these L-asparaginase-treated patients may not be manifest, however, because the system is not stressed. There is destruction of the leukemia cells which sometimes can lead to disseminated intravascular coagulation; however, there is not the massive tissue destruction seen in patients with multiple trauma injuries or severe burns. If these L-asparaginase-treated patients' plasma were put in an in vitro opsonic assay, and titered out it should demonstrate a relative opsonic deficiency compared to normals. Importantly these patients manifest no symptomatology of reticuloendothelial system dysfunction. In vivo, however, these levels appear to have no untoward effects. Therefore, some care seems warranted in interpreting opsonic activity and the relationship between decreased fibronectin and reticuloendothelial blockade.

The second area of concern deals with the interpretation of cryoprecipitate transfusion studies. Certain studies have shown improvement in patient status following cryoprecipitate therapy and asserted that this was due to fibronectin [333]. Patients who have low fibronectin levels may have low fibrinogen levels and low coagulation factor VIII levels because of the disseminated intravascular coagulation that may be occurring. Administration of cryoprecipitate will rectify the low levels of fibronectin and factor VIII, and possibly ameliorate the patient's condition on this basis. It would seem that the specific merits of fibronectin await further testing to determine whether purified fibronectin without other plasma constituents modifies important physiological parameters and improves the patient's clinical status. Preliminary studies have reported that affinity-purified fibronectin will correct reticuloendothelial system depression [385].

Studies from Austin's laboratory examined monocyte phagocytosis of particulate activators of the alternative complement pathway. These particles included zymosan, sheep, rabbit, and mouse erythrocytes, and erythrocytes sensitized with various complement components. Surprisingly, intact fibronectin was not terribly effective in promoting monocyte phagocytosis in these systems [336]. It was observed that a fibronectin component would promote what was termed "opsonin-independent phagocytosis" [336]. In fact it is fibronectin, or, probably more correctly, fibronectin fragments, that were acting as an opsonin differently from traditional immunoglobulin and compliment mechanisms. The studies by Czop et al [336] showed that intact fibronectin did not promote particulate opsonization, but that fragments of fibronectin generated by tryptic degradation and purification on a monoclonal antibody column would promote the opsonization. These studies should be considered when evaluating the previously described opsonization of gelatin-coated lipid emulsions or latex beads. That is, some fragments of fibronectin may promote opsonization while others may block reticuloendothelial function, and trace amounts of fragments may be present which would not necessarily be visible on routine gels, but might require silver staining or autoradiography to visualize.

XVII. BACTERIAL BINDING

Fibronectin has been observed to bind to certain strains of staphylococci [337–339]. The adhesion of bacteria to various surfaces and the role this may play in pathogenesis has received a great deal of interest recently. Since fibronectin serves an important function in biological adhesion phenomena, it was logical to examine fibronectin binding to microorganisms. A common strain of Staphylococcus, Cowan I, is high in protein-A content and will bind well to fibronectin [337–339]. In other studies, a variety of strains and mutants of Staphylococci were examined to determine what specific constituents of the bacterial cell wall were responsible for the binding. The major constituents of the Staphylococcal wall are teichoic acid, peptidoglycan, protein A, and some other quantitatively minor protein constituents [340–342]. Studies by Kuusela and our laboratory, using enzyme-linked assays or binding of radiolabeled fibronectin, showed that the protein A did not appear to be responsible for fibronectin binding [337, 339]. In addition, there was no correlation between protein-A content of the various Staphlococcal strains and the amount of fibronectin bound [339]. Further studies which used variants or mutants with low peptidoglycan or teichoic acid content failed to show any direct association between fibronectin binding and the level of these components in bacterial walls [339]. Experiments determined that fibronectin could bind purified cell walls from Staphylococci and that trypsin treatment of the purified cell wall appeared to degrade the constituent responsible for fibronectin binding. Other investigations showed that encapsulated strains of Staphylococci had very low fibronectin binding. In summary, studies from our laboratory by Verbrugh et al [339] and Kuusela [337] would suggest that fibronectin binds to some protein constituent in the cell wall other than protein A and that the role of fibronectin in opsonization remains uncertain. Preliminary studies from our laboratory have shown that fibronectin will promote adhesion of certain bacteria to endothelial cells and will lead to an aggregation of these microorganisms (unpublished observation).

Fibronectin appears to show very low binding to gram negative organisms, but E coli has been the gram-negative organism most extensively tested [337–339]. Other work has implicated fibronectin in diseases in which gram-negative organisms are involved, such as Pseudomonas in cystic fibrosis patients where it was shown that fibronectin increased adhesion of Pseudomonas to buccal epithelial cells [343]. Other studies have reported that pseudomonas aeruginosa adhesion to buccal cells is associated with a loss of fibronectin from the cell surface which occurs in seriously ill, colonized patients or can be mimicked by trypsin treatment of cells [344]. Adhesion of bacteria to various sites—eg, vaginal, oral mucosa, bladder epithelia, or endothelia in cases of subacute bacterial endocarditis—has become an area of intense interest relative to mechanisms of disease. Specifically, how adhesion of a bacterium initiates the sequence of events leading to tissue invasion and a bacterial infestation remains to be delineated. Obviously, factors other than bacterial adhesion, such as numbers of bacteria, presence of antibody, mechanical defects in tissues, blood supply, etc, must be important. Specifically whether fibronectin can modulate bacterial adhesion and thereby the pathogenesis of bacterial infections remains to be shown. The binding of fibronectin to bacteria, the interactions of bacteria and platelets, which are known to have fibronectin on and within them (see platelet section), provide substantial support for further investigating this area.

XVIII. FUNCTIONAL AND ANTIGENIC DOMAINS

Fibronectin is a globular molecule with biologically active domains held together by flexible polypeptide segments. Limited proteolysis or chemical fragmentation has permitted the isolation and purification of peptide fragments with various biological activities. These domains will not be discussed in the order of discovery, but rather in a combination of what may be the most understandable manner, and a model of the domain organization is seen in Figure 9. The amino terminus of the molecule consists of a 27- to 30-kd fragment that has diverse functions. This fragment can be generated by various proteases including thrombin, trypsin, plasmin, or cathepsin D [among others see 22, 101, 338, 345–347]. This 27-kd amino-terminal segment is the site for binding to Staphylococcus [338], and the transamidation or cross-linking site for fibrin, collagen, and itself [144, 294, 348]. This amino-terminal domain also has weak heparin-binding activity, discussed below [346, 349]. Sequence data have also shown that the amino-terminal residue is blocked by pyroglutamic acid and continues in from the amino terminus Glu-ALA-GLX-GLX-Val for human plasma fibronectin [22]. Substantial progress has been made in sequencing bovine plasma fibronectin, which has shown that there are five repeats of homologous disulfide-looped "fingers" in an amino-terminal 29-kd fragment [350]. Interestingly, this disulfide-looped "finger" is repeated three times in a carboxy-terminal pepsin fragment.

Perhaps one of the best studied "domains" of fibronectin to date is that which binds to gelatin. This has been variously ascribed to 30- to 70-kd tryptic fragments [351–354], a 40-kd chymotrypsin fragment [353, 354], a 40-kd plasmin fragment [345], cathepsin and elastase fragments with molecular weights of 40 kd [355], and subtilisin generated peptides of 30–50 kd [104]. The gelatin-binding domain has also been studied by various other groups [101, 163, 349, 351]. Earlier studies had shown that the collagen-binding site was in the amino terminus of the molecule [345], toward the carboxyl end of the 27-kd amino-terminal peptide [351] (Fig. 9).

A variety of studies have attempted to define the localization of the gelatin-binding site. Studies of Ruoslahti et al [352] isolated fragments from human cryoprecipitate that ranged from 80 to 205 kd. These fragments were then isolated by gelatin-affinity chromatography and studied for other functions including: 1) binding to antibody specific for the 70-kd gelatin binding region; 2) heparin binding; and 3) mediation of cell attachment. These fragments all lacked at least the NH_2 terminus ($\sim$ 27 kd) seen in the intact molecule, which is the Staphylococcus- and fibrinogen-binding site and cross-linking site for collagen discussed above. Ruoslahti et al [352] performed NH_2 terminus amino acid sequencing which permitted the determination of which fragments originated from the same site on the molecule. With this, it was observed that normally occurring fragments of fibronectin in cryoprecipitate or fragments purposefully generated with trypsin generated peptides of 30–200 kd which bound gelatin; with fragments of 80–200 kd all having the same amino-terminal sequence, and therefore all arise from the same amino-terminal site cut by protease [352]. These pieces all have the amino terminus from the native molecule removed; this site seems very sensitive to various proteases including plasmin, trypsin, thrombin, and cathepsin [22, 101, 144, 338, 347]. Studies from our laboratory using tryptic and catheptic fragments of human fibronectin, along with monoclonal antibodies, have shown that the gelatin-binding domain is in a 5-kd piece running from 27 to 32 kd of the amino-terminal end of the intact molecule [346] (see model, Fig. 9). Other studies have shown that this same region of the molecule is near the site for actin binding [356–357], as evidenced by the observation that actin can compete for gelatin binding [358]. These studies proposed that the actin-binding site was in a 27-kd amino-terminal fragment and that the proximity of the actin- and gelatin-binding sites was such that each sterically hindered the binding of the other (see model, Fig. 9). Somewhere near these sites may reside a domain which binds DNA and the physiological relevance of this is unclear [359].

Fibronectin will bind to cells and promote cell attachment activity, and this seems to reside in peptide domains within the molecule [349, 352, 353]. In general, the larger the fragments evaluated, the more additional activities were observed. In the cryoprecipitate fragments of Ruoslahti et al [352], a fragment bound collagen and mediated cell attachment while a 200-kd fragment bound collagen, mediated cell attachment, and bound to heparin. These activities seem all to be originating from unique domains of the molecule. This is similar to observations made by Yamada's group which showed that a $\sim$ 180-kd peptide was the smallest fragment that would bind collagen and

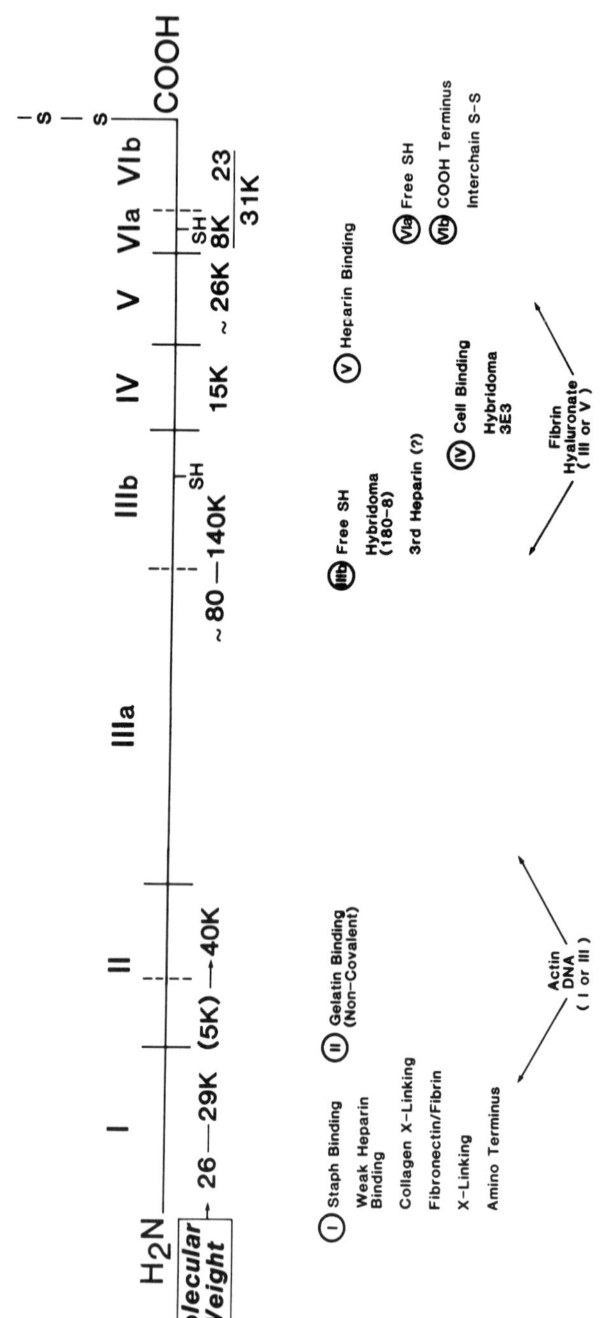

H₂N
COOH
Molecular Weight
I
II
IIIa
IIIb
IV
V
VIa
VIb
26—29K
(5K)
40K
~80—140K
15K
~26K
8K
23
31K
SH
SH
SH
—S—S—
I Staph Binding
   Weak Heparin Binding
   Collagen X-Linking
   Fibronectin/Fibrin
   X-Linking
   Amino Terminus
II Gelatin Binding (Non-Covalent)
IIIb Free SH
   Hybridoma (180-8)
   3rd Heparin (?)
IV Cell Binding
   Hybridoma 3E3
V Heparin Binding
VIa Free SH
VIb COOH Terminus
   Interchain S-S
Actin
DNA
(I or III)
Fibrin
Hyaluronate
(III or V)

mediate cell attachment using chicken fibronectin [353, 360]. Gold et al [104] reported the presence of a 50-kd fragment of fibronectin that mediated weak cell attachment along with gelatin binding. These data are somewhat disparate with the studies cited above. In particular, those studies of Ruoslahti [352], Yamada [354], and our laboratory [346] demonstrated an 80-kd gelatin-binding fragment that must come immediately adjacent to the 27-kd amino-terminal staph/fibrin-binding peptide. That being the case, if the studies of Gold et al [104] were correct, this 80-kd domain should promote cell attachment, and this has not been observed (see model, Fig. 9). Possibly, trace contamination of different fragment preparations which are enriched predominantly for one could account for the disparate observations. The studies of Gold et al [104] aside, other studies support the position of the gelatin and amino-terminal Staphylococcal binding domain [101, 351]. Wagner and Hynes used tryptic fragments and cyanide cleavage to determine where various biological or chemical activities were located. The conclusions of Wagner and Hynes [163, 257] among others were that 1) a 25-kd fragment (analogous to 27 kd) was generated from the amino terminus; 2) interchain disulfide bonds are within 10 kd of carboxyl terminus of the molecule; 3) that one or two cysteines were present per chain (see other section on free $-$SH in fibronectin); 4) the amino terminus of the molecule has a large number of intrachain disulfide bonds. Cyanide cleavage studies have also been performed by Sekiguichi et al [21]. These studies are consistent with

---

Fig. 9. Schematic representation of a model of fibronectin broken down into domains with Roman numerals and approximate molecular weights listed below. This is a composite model from the work of various investigators' efforts which are cited in the text. Domain I, of 26–29 kd, is the amino-terminal domain with the following activities: Staph binding, weak heparin-binding and cross-linking sites for collagen, fibronectin, and fibrin. Immediately adjacent to this region is the noncovalent gelatin binding region, domain II. This has been reported to be in fragments of approximately 40 kd, and by inference we locate it in a 5-kd region immediately adjacent to the amino-terminal 27-kd domain I [346, 347]. Actin- and DNA-binding activities have been reported, and these may reside in domain I or III or both [reviewed in 360]. Region III of fibronectin is a large segment in the middle of the molecule without many known functions. It has been shown, however, that a fragment at the COOH terminal side, IIIB, possesses one of the two free SH in fibronectin [347], binds a hybridoma designated 180-8 [346, 347], and may be a third heparin-binding domain [360]. Domain IV has cell-binding activity and has been most precisely defined with a monoclonal antibody, 3E3, by Piersbacher et al [367] and has a molecular weight of 15 kd. Adjacent to this is domain V which is a strong heparin-binding domain with a molecular weight of 26 kd or greater depending on which potease is used. Other activities, including fibrin and hyaluronate binding, are in domain III or V. The possible differences in the two chains of plasma fibronectin would possibly be in domain III or V. At the COOH terminus of the molecule is a 31-kd peptide which possesses a second free SH in VIa and has the interchain disulfide bonds (VIb). Studies have radiolabeled the SH in the 31-kd fragment which is then degraded to an 8-kd labeled piece and a 23-kd unlabeled piece [346, 347].

studies of Furie et al [251], who used thrombin and cathepsin D 1) to cleave off an amino-terminal 29-kd thrombin-sensitive piece, and 2) to isolate an adjacent 43-kd gelatin-binding region.

XIX. HEPARIN/PROTEOGLYCAN BINDING

A number of laboratories have evaluated the binding of fibronectin to heparin as an indicator of proteoglycan interaction (see above). Peptide fragments of plasma fibronectin which bind heparin have recently been identified. Studies by Balian et al [345] showed the presence of an NH_2-terminal fragment of about 70 kd generated by cathepsin D which bound weakly to heparin. This peptide and larger fragments also had gelatin-binding activity [345], and contained the site for transamidation by coagulation factor XIII_a which placed them near the NH_2 terminus [361]. The heparin-binding "domain" was described to reside in a 30-kd piece generated from catheptic digests [345, 346]. Tryptic digestion will release a 27-kd amino-terminal peptide (discussed above), which also shows weak heparin-binding activity [346, 349]. It has recently been shown that this amino-terminal heparin interaction is inhibited by divalent cations.

Another, somewhat stronger heparin-binding site on the molecule has been described [346, 349, 361–364]. Studies by Richter et al [361] determined that this heparin-binding fragment of 140 kd on nonreduced gels consisted of two disulfide-bonded peptides that migrated slightly differently on reduced gels with molecular weights of 65 and 75 kd. These two peptides were presumably generated from different subunits of the fibronectin dimer. Also, the larger fragment contained an additional transamidation site (distinct from that present in the 27-kd NH_2-terminal fragment) while the 65 kd did not. The conclusion drawn was that the larger fragment contained an additional peptide sequence [361]. Also, the transamidation site may be in a fragment analagous to a 21-kd thermolysin fragment isolated from the alpha (the larger) subunit of plasma fibronectin [365].

Using heparin sepharose and monoclonal antibody affinity chromatography, we have determined the location of this additional heparin-binding domain. Clearly, as cited above, there is a weak heparin-binding domain at the amino terminus which can be generated by trypsin, cathepsin, plasmin, or thrombin cleavages. At a minimum, a second heparin-binding domain exists in the carboxy third of the molecule. Using peptide mapping, monoclonal antibodies, and heparin or gelatin chromatography and focusing on a heparin-binding domain in the carboxy terminal region of the molecule, the following observations were made [346]. These studies show that there is a 31-kd carboxy-terminal tryptic peptide (which presumably has had the interchain S-S removed at the extreme COOH terminus) [351] which does not bind heparin, does bind a monoclonal antibody designated 2-8, and contains a free sulfhydryl discussed in detail below [346, 347]. This carboxy-terminal region of the molecule contains an intrachain disulfide-bonded loop or amino acid sequence which contains three reiterative sequences homologous to a fivefold reiteration of similar sequences in the 27-kd amino-terminal domain [350]. Cathepsin D fragments generated by successive cleavages from the COOH terminus have molecular weights of 66, 43, 33, and 29 kd, bind monoclonal antibody 2-8 and heparin, whereas monoclonal 2-8 reacts with a non-heparin-binding fragment at the extreme COOH terminus (see Fig. 9). Therefore monoclonal 2-8 has an antigenic determinant in the 31-kd COOH-terminal peptide [346]. Immediately adjacent to this 31-kd peptide running from its NH_2 terminus toward the middle of the molecule is a second heparin-binding region seen in its smallest component of 29 kd. This fragment appears adjacent to and possibly incorporates a portion of NH_2 terminus of the COOH terminal 31-kd piece (Fig. 9). This fragment is probably analogous to heparin-binding fragments seen by other investigators which do not originate from the amino terminus of the molecule [362]. Studies from Yamada's laboratory have shown further that in this region of the molecule is another site which will bind hyaluronate [362]. The binding of hyaluronate does not compete with that for heparin, and therefore presumably represents a unique "domain." Recent studies by Laterra and Culp [366] have shown that plasma fibronectin failed to bind hyaluronate under various conditions. Cellular-derived fibronectin will bind hyaluronate [362], but only under special conditions [366]. Specifically, the binding of cellular fibronectin to hyaluronate seems to be contingent upon multimeric aggregation of cellular fibronectin [366].

In this same region of the molecule appears to be a domain responsible for cell attachment or cell adhesion. Studies from a variety of laboratories have shown that proteolytic fragments of fibronectin would promote cell adhesion (see citations above). Most recently, studies from Ruoslahti's laboratory have isolated a monoclonal antibody which reacts with the region of the molecule responsible for cell binding [367]. This cell-binding domain has further been refined to a 15-kd region of the molecule. The possible topographical location of this site is seen in Figure 9. It is not certain what the specific location of this region is relative to the carboxy heparin-binding or hyaluronate-binding region. The most probable position of the cell binding domain is more toward the NH_2 terminus of the molecule than the COOH heparin domain and certainly in 31 kd from the COOH terminus of the intact molecule (Fig. 9).

Fibronectin has also been shown to bind to fibrin or fibrinogen, as discussed above. Recent studies have evaluated which peptide domain of fibronectin binds to fibrin. Sekiguichi and Hakomori [368] have shown that thermolysin generated a 24-kd and 21-kd fragment which bound to heparin. These investigators proposed two unique heparin-binding sites by comparing tryptic and thermolysin digests. In examining the 200- and 180-kd tryptic fragments, it was observed that a 21-kd fibrin-binding fragment was generated only from the 200-kd precursor, and that this originated at or near the COOH terminus. It was further proposed that there was a separate 24-kd thermolysin product, similar to a 32-kd amino-terminal tryptic peptide (27 kd as defined by others) which reacted with fibrin (Fig. 9). It was proposed by Sekiguichi and Hakomori [368] that there is an asymmetric presence of a 21-kd peptide domain near the COOH end of the molecule, it being apparently only in the higher molecular weight subunit. One possible concern regarding these studies is whether the 32-kd amino-terminal fibrin-binding domain can be further cleaved to not only 24 kd, but also 21 kd to account for the above data. Further studies should be able to define this. In support of this contention of an additional domain in one chain are studies of Richter et al [361], who showed a heparin-binding domain, near the COOH terminus, that had 75-kd and 65-kd apparent molecular weights, the larger also having a transamidation site. By contrast, our studies with monoclonal antibodies recognizing certain heparin-binding peptides at the COOH terminus do not necessarily support the conclusion of an extra peptide segment in this region of the molecule, but this is certainly possible [246].

XX. SULFHYDRYL-DISULFIDE STATUS OF FIBRONECTIN

Fibronectin contains 20–25 disulfide bridges. Most are intrachain and are concentrated in the amino-terminal third of the molecule [5, 101, 163, 345, 355]. The interchain disulfide bridge(s) responsible for linking together dimers are at the carboxyl terminus [22, 350]. In addition, free sulfhydryls have been detected in soluble fibronectin; however, these cannot be titrated in the absence of denaturant [15, 247, 355]. After denaturation in sodium dodecyl sulfate, approximately 0.9 moles or more (see below) of cysteine per 200 to 220 kd subunit can be titrated [15, 247, 355]. The pattern of cleavage with cyanide suggests one or two free sulfyhydryls per subunit [163, 369]. Cyanide cleavage experiments [163, 355, 369] and analysis of elastase fragments position the free sulfhydryl(s) toward the carboxyl terminus [355].

Insoluble matrix fibronectin exists in disulfide-bonded oligomers as well as dimers [370, 371]. The sulfhydryl status of fibronectin seems important because oxidation of free sulfhydryls [134, 372], and blockage or modification of free sulfhydryls, prevent binding of soluble fibronectin to the cell layer [101].

Recently, we have used monoclonal antibodies to define the location of free sulfhydryl(s) in plasma fibronectin [247]. These data showed that there are 1.3–1.6 free sulfhydryls per 220 kd monomeric subunit and that these could be titrated only under denaturing conditions—for example, 3 M guanidine or 6.3 M urea. By two different methods, reasonably equivalent quantities of free sulfhydryl were observed (Table II). Monoclonal antibody 2-8

TABLE II. Titration of Fibronectin Sulfhydryls With DTNB and 2-PDS

Denaturant	Titrant	No. of experiments	Moles/2.0 × 10⁵ g protein	
			Mean	SD
3 M Guanidine	DTNB	5	1.6	± 0.1
	2-PDS	6	1.3	± 0.1
6.3 M Urea	DTNB	5	1.5	± 0.1
	2-PDS	5	1.4	± 0.1

Titration of fibronectin sulfhydryls with DTNB (5,5′-dithiobis) (2-nitrobenzoic acid) or 2-PDS (2,2′-dipyridyl disulfide or 2,2′-dithiopyridine) [247].

recognized a 31-kd, free SH-containing peptide generated in early tryptic digests which is located near the carboxyl terminus (Fig. 9). A second free sulfhydryl was observed in an 80-kd tryptic peptide which reacted with hybridoma 180-8 but not hybridoma 2-8 [247]. Tryptic fragments were purified and titrated individually with 5,5′-dithiobis (2-nitrobenzoic acid) in 4.0 M guanidine, which suggested separate free SHs in the 31-k and 80-k tryptic fragments. Both the 31-kd and 80-kd tryptic peptides bind to thiol-Sepharose under denaturing conditions. Finally, free sulfhydryls were labeled in intact fibronectin with ^{3}H-iodoacetic acid in 3.0 M guanidine [247]. When the S-^{3}H-carboxymethylated fibronectin was cleaved, the 31-kd labeled peptide at the COOH terminus was degraded to an unlabeled 23 kd peptide which reacted with hybridoma 2-8 leaving the ^{3}H-carboxy-methyl label by inference, in an approximate 8-kd fragment within this 31-kd piece. This leads to the following presumptive model regarding the location of free SH in fibronectin (Fig. 9). It is important to note, however, that these sites are buried and not readily accessible without denaturants.

It is likely that the covalent association of fibronectin with itself or other molecules may be important in the oligomer formation of fibronectin and possibly mediate functional signals to cells. It would seem that this will be a fruitful area for further investigation.

XXI. FIBRONECTIN AND DISEASES

Plasma fibronectin has been reported to be decreased in major trauma, major surgery and burns [see 322, and references therein]. Levels of fibronectin were studied by Mosher and Williams [290] in severely ill patients and were found to be decreased in a variety of patients. These low levels seemed to be correlated with disseminated intravascular coagulation [290].

Other situations with low plasma fibronectin levels are thought to be due to excess consumption, perhaps as a nonspecific opsonin (see opsonin section). Some of the earliest studies on plasma fibronectin showed that it could be elevated in certain patients with cryofibrinogenemia [14]. Other more recent studies reported that fibronectin is a component of cryoglobulins, which have been studied in patients with various connective tissue and other nonrelated diseases [373]. There are reports that fibronectin levels are increased in patients with connective tissue diseases [373]. Studies by Stathakis et al [329] measured fibronectin by Laurell rocket immunoassay in more than 100 controls and 149 patients with various diseases. This studies showed that normal men had slightly higher levels (339 ± 76 µg/ml) than normal women (311 ± 82 µg/ml), and thus corroborated earlier studies by Mosher and Williams [290]. Stathakis et al [329] found that levels of fibronectin were increased in patients with biliary cirrhosis, obstructive jaundice, and nephrotic syndrome; had no consistent pattern in advanced cancer, which was previously determined by Mosher and Williams [290]; and were decreased in cancer patients with cryofibrinogenemia or with extensive liver metastasis. These last studies could be interpreted to suggest that the liver is a significant source for the production of plasma fibronectin.

Fibronectin has also been implicated in a variety of disease processes, or found in cells that are associated with various diseases. Some representative examples follow. Fibronectin was localized by immunohistochemical techniques in vessel walls and found to be prominent in atherosclerotic lesions of the intima, particularly in what appeared to be developing plaques [374].

Fibronectin has been measured in various body fluids besides blood, but only limited associations with disease have been made. Fibronectin was reported in urine, and found to be increased in urine in cases of prostatic cancer [40]. These last studies measured protein binding to gelatin columns and presumed this was fibronectin. It would be useful and more specific to perform studies by ELISA or radioimmunoassay. Also, the studies of Webb and Linn [40] measuring fibronectin in urine in prostatic cancer relied on quantitation by gelatin binding which potentially could be measuring fragments or possibly plasma fibronectin that may have leaked into urine along with hematuria, as might occur with an invasive carcinoma.

Interesting studies by Clark et al [375] examined fibronectin in delayed hypersensitivity reactions. The studies utilized delayed hypersensitivity to PPD in guinea pigs and measured 1) accumulation of ^{125}I-fibronectin or albumin at these sites, and 2) de novo synthesis of fibronectin at these sites. These studies reported that the microvasculature regulates the accumulation of fibronectin by changes in increased vascular permeability and also by synthesis at local sites [375] perhaps where endothelial cells are proliferating. Various studies cited elsewhere in the paper deal with the role of fibronectin as an opsonin and modulation of macrophage function. Possibly related to the reticuloendothelial system are studies by Resnick and Nachman [376], which showed that Reed-Sternberg cells, which are pathognomonic of Hodgkin disease, contain fibronectin. Based on this observation, Reed-Sternberg cells were suggested to be of macrophage origin, not of endothelial or fibroblast origin [376].

Increased amounts of fibronectin were observed in the glomerulus in membranoproliferatrive glomerulonephritis [46, 377], though Scheinman et al [47] did not make similar observations. Other studies showed fibronectin to be absent from hyalinized end-stage glomeruli and old segmental scars [17, 49]. Fibronectin appears to be a component of the reaction to injury in the early stages of glomerulonephritis. The source of the fibronectin may be local production by cells in the mesangium or transudation of plasma fibronectin during the inflammatory reaction in the glomerulus. The reason for the absence of fibronectin in "old," end-stage renal lesions is unknown, and what the molecular constituents of these lesions are awaits further study, but abnormal accumulation of basement membrane components is a distinct possibility. Fibronectin is also substantially increased in the mesangium of the diabetic kidney [47].

Fibronectin and other basement membrane constituents have been studied in other diseases such as liver fibrosis [378]. In liver fibrosis there was observed an intense reaction with antifibronectin antibodies reacting with large fibrotic septa and periportal interstitial regions. Also reported were certain unique distributions for fibronectin and laminin in the fibrotic liver [378].

Fibronectin may participate in modulating reparative processes which may become excessive and lead to extensive fibrosis. Other sections described in this chapter deal with the chemotactic stimuli generated by fibronectin for fibroblasts, monocytes, and other cell types. Additionally, fibronectin may serve as a substratum for cell adhesion and movement, thereby also directing cells to a specific site.

A study by Rennard and Crystal [379] quantitated the production of fibronectin by alveolar macrophages from normal individuals and patients with idiopathic pulmonary fibrosis and sarcoidosis, among other fibrotic lung diseases. This study showed excessive production of fibronectin by alveolar macrophages from patients with fibrotic lung diseases. It was further shown that this macrophage produced fibronectin was a chemoattractant for fibroblasts, and hypothesized that the excess production of fibronectin attracted fibroblast cells to various sites and this led to the development of the fibrotic lung disease [379]. It is also of note that levels of fibronectin have been observed to be increased in bronchial lung fluid of cigarette smokers compared with nonsmokers [380].

The studies in section III cite the presence of fibronectin in skin of lower vertebrates, mammals, and man. Fayand has shown that areas of skin affected by psoriasis have staining for fibronectin in the epidermis, largely in the cornified layers, with other changes occurring in the dermis and basement membrane [54]. Other studies have shown fibronectin to be increased in the dermis in scleroderma [381]. Plasma levels of fibronectin, however, have not been shown to be increased in scleroderma, a disease with generalized excessive accumulation of collagen.

Other studies measured fibronectin in the synovial fluid where the levels were 445 ± 103 μg/ml (mean + SD), which was higher than the corresponding mean plasma levels in these patients [39]. Because limited synovial fluid is present under normal conditions, it was impossible to precisely compare the synovial fluid fibronectin level in rheumatoid patients to that of normals. In rheumatoid arthritis, fibronectin is present in increased amounts in the synovium and is deposited with fibrin during fibrosis [383]. In addition, the levels of fibronectin in synovial fluid is greater than that of plasma in cases of rheumatoid arthritis and various other arthritic conditions [383]. Importantly, however, the levels of plasma fibronectin in these various arthritic conditions are not increased above normal [373].

Interestingly, in various fibrotic conditions associated with human disease—eg, cystic fibrosis, nodular sclerosing Hodgkin disease, and scirrhous carcinoma of the breast—fibronectin is a major constituent of the process (unpublished observation). The interrelationship of fibronectin and collagen in these diseases may prove useful in modulating abnormal fibrosis that often occurs as a component of many pathological processes.

Finally, it has been shown that adhesion of late passage representing "old" or aged fibroblasts is defective compared to normals [384]. However, fibronectin from early passage cells promotes maximal adhesion of both young and old cells [384]. What role fibronectin or other matrix constituents play in cellular aging remains to be defined.

ACKNOWLEDGMENTS

I wish to thank the following colleagues for their helpful suggestions, stimulating conversations, and provocative questions: at the University of Minnesota, Michael Berry, Arianna Chuang, Dr. Jim McCarthy, Sally Palm, Dr. Michael Silver, and Dr. Dennis Smith; and at other institutions, Drs. Hynda Kleinman, George Martin, Deane Mosher, Erkki Ruoslahti, and Ken Yamada, among others. I also thank Patricia A. Lopez, Carol Lynn Larson, and Linda Samide-Kenny for their assistance in the preparation of this manuscript. The author is the recipient of the Stone Professorship of Pathology from the University of Minnesota and a research career development award NIH/NCI CA-00651. This work was supported by grants CA 21463 and CA

29995 from the NIH, and grants from the Leukemia Task Force, University of Minnesota.

XXII. REFERENCES

1. Morrison PR, Edsall JT, Miller SG: J Am Chem Soc 70:3103, 1948.
2. Vaheri A, Mosher DF: Biochim Biophys Acta 516:1–25, 1978.
3. Yamada K, Olden K: Nature 275:179–184, 1978.
4. Mosesson MW: Throm Haemostas 38:742–750, 1977.
5. Pearlstein E, Gold LI, Garcia-Pardo A: Mol Cell Biochem 29:103–128, 1980.
6. Ruoslahti E, Engvall E, Hayman EG: Coll Res 1:95–128, 1981.
7. Mosher DF, Furcht LT: J Invest Dermatol 77:175–180, 1981.
8. Hynes, RO: Biochim Biophys Acta 458:73–101, 1976.
9. Kleinman H, Klebe R, Martin GR: J Cell Biol 88:473–485, 1981.
10. Edsall JT, Gilbert GA, Scherage HA: J Am Chem Soc 77:157, 1955.
11. Smith RT, Von Korff R: J Clin Invest 36:596–604, 1957.
12. Mosesson MW, Colman RW, Sherry S: N Engl J Med 278:815–821, 1968.
13. Mosesson MW, Umfleet RA: J Biol Chem 245:5728–5736, 1970.
14. Mosesson MW, Chen AB, Huseby RM: Biochim Biophys Acta 386:509–524, 1975.
15. Mosher DF: J Biol Chem 250:6614–6621, 1975.
16. Alexander SS, Colonna G, Edelhoch H: J Biol Chem 254:1501–1505, 1979.
17. Yamada K, Weston JA: Proc Natl Acad Sci USA 71:3492–3496, 1974.
18. Bray BA, Mandl I, Turino GM: Science 214:793–795, 1981.
19. Kurkinen M, Vartio T, Vaheri A: Biochim Biophys Acta 624:490–498, 1980.
20. Fukuda M, Hakomori S: J Biol Chem 254(12):5442–5450, 1979.
21. Sekeguchi K, Fukuda M, Hakomori S: J Biol Chem 256(12):6452–6462, 1981.
22. Furie MB, Rifkin DB: J Biol Chem 255:3134–3140, 1980.
23. Jaffe EA, Mosher DF: J Exp Med 147:1779–1791, 1978.
24. Crouch E, Balian G, Holbrook K, Duksin D, Bornstein P: J Cell Biol 78:701–715, 1978.
25. Chen AB, Amramni DL, Mosesson MW: Biochim Biophys Acta 493:310–322, 1977.
26. Iwanaga S, Suzuki K, Hashimoto S: Ann NY Acad Sci 212:56–73, 1978.
27. Yamada K, Kennedy DW: J Cell Biol 80:492–498, 1979.
28. Vuento M, Wrann M, Ruoslahti E: FEBS Lett 82:227–231, 1977.
29. Engvall E, Ruoslahti E: Int J Cancer 20:1–5, 1977.
30. Dessau W, Jilek F, Adelmann BC, Hörmann H: Biochim Biophys Acta 533:227–237, 1978.
31. Ruoslahti E, Vuento M, Engvall E: Biochim Biophys Acta 534(2):210–218, 1978.
32. Vuento M, Vaheri A: J Biochem 175:333–336, 1978.
33. Klebe RJ, Bentley KL, Sasser PJ, Schoen RC: Exp Cell Res 130:111–117, 1981.
34. Molnar J, Gelder FB, Zong LM, Siefring GE, Credo RB, Lorano L: Biochemistry 18:3909–3916, 1977.
35. Chen AB, Mosesson MW, Solish G: Am J Obstet Gynecol 125:958–961, 1976.
36. Alitalo K, Kurkinen M, Vaheri A, Krieg T, Timpl R: Cell 19:1053–1062, 1980.
37. Ruoslahti E, Engvall E, Hayman E, Spiro R: Biochem J 193:295–299, 1981.
38. Kuusela P, Vaheri A, Paolo J, Ruoslahti E: J Lab Clin Med 94:595–601, 1979.
39. Vartio T, Vaheri A, Von Essen R, Isomaki H, Stenman S: Eur J Clin Invest 11:207–212, 1981.
40. Webb SK, Lin HG: Invest Urol 17:401–404, 1980.
41. Linder E, Vaheri A, Ruoslahti E, Wartiovaara J: J Exp Med 142:41–49, 1975.
42. Quaroni A, Isselbacher KJ, Ruoslahti E: Proc Natl Acad Sci USA 75:5548–5552, 1978.

43. Stenman S, Vaheri A: J Exp Med 147:1054–1064, 1978.
44. Repesch L, Furcht LT, Smith DE: J Histochem Cytochem 29:937–945, 1981.
45. Wartiovaara J, Stenman S, Vaheri A: Differentiation 5:85–89, 1976.
46. Pettersson EE, Colvin RB: Clin Immunol Immunopathol 11:425–436, 1978.
47. Scheinman JI, Fish JA, Matas JA, Michael FA: Am J Pathol 90:71–84, 1978.
48. Oberley DT, Mosher DF, Matthew DM: Am J Pathol 96:651–662, 1979.
49. Dixon AJ, Burns J, Dunnill MD, McGee J: J Clin Pathol 33:1021–1028, 1980.
50. Courtoy JP, Kanwar SY, Hynes OR, Farquhar GM: J Cell Biol 87:691–696, 1980.
51. Burns J, Dixon AJ, Woods JC: Histochemistry 67:73–78, 1980.
52. Madri AJ, Roll J, Furthmayr H, Foidart MJ: J Cell Biol 86:682–687, 1980.
53. Kanwar YS, Farquar YS: Proc Natl Acad Sci USA 76:1303–1307, 1979.
54. Fryand O: Arch Dermatol Res 266:33–41, 1979.
55. Hynes RO, Martin GS, Critchley DR, Shearer M, Epstein CJ: Dev Biol 48:35–46, 1976.
56. Furcht LT, Mosher DF, Wendelschafer-Crabb G: Cell 13:261–271, 1978.
57. Chen LB: Cell 10:393–400, 1977.
58. Hassell JR, Pennypacker JP, Yamada K, Pratt RM: Ann NY Acad Sci 312:406–409, 1978.
59. Vaheri A, Ruoslahti E, Westermark B, Ponten J: J Exp Med 143:64–72, 1976.
60. Schater M, Schoonmaker G, Hynes RO: Brain Res 159:149–158, 1978.
61. Kurkinen M, Alitalo K: FEBS Lett 102:64–68, 1979.
62. Wilson BS, Ruberto G, Ferrone S: Biochem Biophys Res Commun 101:1047–1051, 1981.
63. Macarak EJ, Kirby E, Kirk T, Kefalides NA: Proc Natl Acad Sci USA 75:2621–2625, 1978.
64. Birdwell CR, Gospodarowicz D, Nicolson GL: Proc Natl Acad Sci USA 75:3273–3277, 1978.
65. Vlodavsky I, Johnson LK, Greenburg G, Gospodarowicz D: J Cell Biol 83:468–486, 1979.
66. Sage H, Crouch E, Bornstein P: Biochemistry 18:5433–5422, 1979.
67. Goldminz D, Vlodavasky I, Johnson LK, Gospodarowicz D: Exp Eye Res 29:331–351, 1979.
68. Mosher DF, Saksela D, Keski-Oja J, Vaheri A: J Supramol Struct 6:551–557, 1977.
69. Killen P, Striker G: Proc Natl Acad Sci USA 76:3518–3522, 1979.
70. Voss B, Allam S, Rauterberg J: Biochim Biophys Acta 90:1348–1354, 1979.
71. Chen LB, Maitland PH, Gallimore PH, McDougall JK: Exp Cell Res 106:39–46, 1977.
72. Smith HS, Riggs JL, Mosesson MW: Cancer Res 39:4138–4144, 1979.
73. Bensusan HB, Koh TL, Henry KG, Murray BA, Culp LA: Proc Natl Acad Sci USA 75:5864–5868, 1978.
74. Plow EF, Birdwell C, Ginsberg MH: J Clin Invest 63:540–563, 1979.
75. Santoro SA, Cunninghm LW: Proc Natl Acad Sci USA 76:2644–2648, 1979.
76. Zucker MB, Mosesson MW, Broekman MJ, Kaplan KL: Blood 54:8–12, 1979.
77. Arnesson MA, Hammerschmidt DE, Furcht LT, Kins RA: JAMA 244:144–147, 1980.
78. Ginsberg MH, Pointer RG, Forsyth J, Birdwell C, Plow EF: Proc Natl Sci USA 77:1049–1053, 1980.
79. Mosher DF, Furcht LT, Wendelschaffer-Crabb G: In Mann KG, Taylor FB Jr (eds): "Regulation of Blood Coagulation." New York: Elsevier North-Holland, 1980, pp 283–289.
80. Johansson S, Rubin K, Hook M, Ahlgren T, Seljelin R: FEBS Lett 105:313–316, 1979.
81. Alitalo KT, Hovi T, Vaheri A: J Exp Med 151:602–613, 1980.
82. Bevilacqua MP, Amrani D, Mosesson MW, Bianco C: J Exp Med 153:42–60, 1981.
83. Van de Water L, Schroeder E, Crenshaw B, Hynes R: J Cell Biol 90:32–39, 1981.
84. Koehler JK, Nudelman ED, Hakomori S: J Cell Biol 86:529–536, 1980.
85. Yamada K, Schlesinger DH, Kennedy DW, Pastan I: Biochemistry 16:5552–5559, 1977.

86. Wrann M: Biochem Biophys Res Commun 84:269–274, 1978.
87. Takasaki S, Yamashita K, Suzuki K, Iwanaga S, Kobata A: J Biol Chem 254:8548–8553, 1979.
88. Fukuda M, Hakomori S: J Biol Chem 254:5451–5457, 1979.
89. Carter WP, Hakomori S: Biochem J 18:730–731, 1979.
90. Olden K, Pratt RM, Yamada K: Proc Natl Acad Sci USA 76:3343–3347, 1979.
91. Hassell JR, Pennypacker JP, Kleinman HK, Pratt RM, Yamada KM: Cell 17:821–826, 1979.
92. Ruoslahti E, Engvall E, Hayman EG, Spiro RG: Biochem J 193(1):295–299, 1981.
93. Teng MH, Rifkin D: J Cell Biol 80:784–791, 1979.
94. Ruoslahti E, Vaheri A, Kuusela P, Linder E: Biochim Biophys Acta 322:352–358, 1973.
95. Ruoslahti E, Vaheri A: J Exp Med 141:97–105, 1975.
96. Keski-Oja I, Mosher DF, Vaheri A: Biochim Biophys Acta 74:699–706, 1977.
97. Yamada K, Schlesinger DH, Kennedy DW: Biochemistry 16:5552–5559, 1977.
98. Kuusela P, Ruoslahti E, Vaheri A: Biochim Biophys Acta 397:295–303, 1975.
99. Chiquet M, Puri EC, Turner DC: J Biol Chem 254:5475–5482, 1979.
100. Yamada K, Kennedy DW: J Cell Biol 80:429–498, 1979.
101. Wagner DD, Hynes RO: J Biol Chem 254:6746–6754, 1979.
102. Matsuda M, Saida T, Hasegawa R: Thromb Res 10:541–546, 1976.
103. Blumenstock FA, Saba TM, Weber P, et al: J Biol Chem 253:4287–4291, 1978.
104. Gold LT, Garcia Pardo A, Frangione B, et al: Proc Natl Acad Sci USA 76:4803–4807, 1979.
105. Yamada K, Yamada S, Pastan I: J Cell Biol 74:649–654, 1977.
106. Hayman EG, Ruoslahti E: J Cell Biol 83:255–259, 1979.
107. Oh E, Pierschbacher M, Ruoslahti E: Proc Natl Acad Sci USA 78(5):3218–3221, 1981.
108. Atherton BT, Hynes RO: Cell 25:133–141, 1981.
109. Hayashi M, Yamada K: J Biol Chem 256:11292–11300, 1981.
110. Bornstein P, Ash JF: Proc Natl Acad Sci USA 74(6):2480–2484, 1977.
111. Hedman K, Vaheri A, Wartiovaara J: J Cell Biol 76:748–760, 1978.
112. Furcht LT, Mosher DF, Wendelschafer-Crabb G: Cancer Res 38:4618–4623, 1978.
113. Furcht LT, Smith D, Wendelschafer-Crabb G, Mosher DF, Foidart JM: J Histochem Cytochem 28:1319–1333, 1979.
114. Furcht LT, Wendelschafer-Crabb G, Mosher DF, Foidart JM: J Supramol Struct 13:15–33, 1980.
115. Culp LA, Murray BA, Rollins BJ: J Supramol Struct 11:401–427, 1979.
116. Rollin BJ, Culp LA: Biochemistry 28:141–148, 1979.
117. Terry AH, Culp LA: Biochemistry 13:414–425, 1974.
118. Furcht LT, Wendelschafer-Crab G, Woodbridge P: J Supramol Struct 7:307–322, 1977.
119. Singer I: Cell 16:675–685, 1979.
120. Peterkovsky B: Arch Biochem Biophys 152:318–328, 1972.
121. Kleinman HK, McGoodwin E, Martin GR, Klebe RJ, Fietzek PP, Woolley D: J Biol Chem 253:5643–5646, 1978.
122. Goldberg B, Green H: J Cell Biol 22:225–258, 1964.
123. Low FN: Anat Rec 142:93–108, 1962.
124. Ruoslahti E, Sjogren HO: Int J Cancer 18:375–378, 1976.
125. Grinnell F: Int Rev Cytol 53:65–144, 1968.
126. Ehrmann RL, Gey GO: J Natl Cancer Inst 16(6):2, 1956.
127. Hauschka SD, White NK: Excerpta Med 240:53–71, 1972.
128. Klebe RJ: Nature 250:248–251, 1974.
129. Rubin KA, Oldenberg A, Hook M, Obrink B: Exp Cell Res 117:165–177, 1978.
130. Murray JC, Liotta LA, Rennard SI, Martin GR: Cancer Res 40:347–351, 1980.
131. Linsenmayer TF, Gibney D, Toole R, Gross J: Exp Cell Res 116:470–476, 1978.

132. Kleinman HK, Murray JC, McGoodwin ER, Martin GR: J Invest Dermatol 71:9–11, 1978.
133. Hewitt AT, Kleinman K, Penneypacker J, Martin G: Proc Natl Acad Sci USA 77:385–388, 1980.
134. Grinnell F, Feld M: Cell 17:117–120, 1979.
135. Ali IU, Mautner V, Lanza R, Hynes RO: Cell 11:115–126, 1977.
136. Yamada K, Yamada S, Pastan I: Proc Natl Acad Sci USA 73:1217–1221, 1976.
137. Grinnell F, Minter D: Biochim Biophys Acta 550:92–99, 1979.
138. Gold LI, Pearlstein E: Biochim Biophys Acta 581:237–251, 1979.
139. Birchmeier C, Kreis TE, Eppenberger HM, Winterhalter KH, Birchmeier W: Proc Natl Acad Sci USA 77:4108–4112, 1980.
140. Chen W-T, Singer SJ: Proc Natl Acad Sci USA 77:7318–7322, 1980.
141. Fox CH, Cottler-Fox MH, Yamada K: Exp Cell Res 130:377–481, 1980.
142. Avnur Z, Geiger B: Cell 25:121–132, 1981.
143. Engvall E, Ruoslahti E, Miller EJ: J Exp Med 147:1584–1595, 1978.
144. Jilek F, Hormann H, Hoppe-Seyler S: Physiol Chem 359:247–250, 1978.
145. Dessau W, Adelmann BC, Timpl R, Martin GR: Biochem J 169:55–59, 1978.
146. Kleinman HK, McGoodwin EB, Klebe RJ: Biochim Biophys Acta 72:426–432, 1976.
147. Jilek F, Hormann H: Hoppe-Seyler's Physiol Chem 360:597–603, 1979.
148. Johansson S, Hook M: Biochem J 187:521–524, 1980.
149. Yamada K, Kennedy DW, Kimata K, Pratt RM: J Biol Chem 255:6055–6063, 1980.
150. Ruoslahti E, Engvall E: Biochim Biophys Acta 631:350–358, 1980.
151. Lindahl U, Hook M: Ann Rev Biochem 47:385–417, 1978.
152. Vuento M, Salonen E, Osterlund K, Stenman U: Biochem J 201:1–8, 1982.
153. Gold LI, Pearlstein E: Biochem J 186:551–554, 1980.
154. Engvall E, Bell ML, Ruoslahti E: Collagen Res 1:505–516, 1981.
155. Vuento M, Salonen E, Salminen K, Pasanen M, Stenman UH: Biochem J 191:719–727, 1980.
156. Kleinman HK, Martin GR, Fishman P: Proc Natl Acad Sci USA 76:3367–3371, 1979.
157. Lustig F: Proc Soc Exp Biol Med 133:207–211, 1970.
158. Duskin DA, Moaz A, Fuchs S: Cell 5:83, 1975.
159. Faulk WP, Conochie LB, Temple A, Papamichail M: Nature 256:123, 1975.
160. Lichtenstein JR, Bauer E, Hoyt R, Wedner H: J Exp Med 44:145–154, 1976.
161. Kjellen L, Pettersson I, Hook M: Proc Natl Acad Sci USA 78:5371–5375, 1981.
162. Mosher DF, Vaheri A: Exp Cell Res 112:323–334, 1978.
163. Wagner DD, Hynes RO: J Biol Chem 255:4304–4312, 1980.
164. Keski-Oja J, Mosher DF, Vaheri A: Cell 9:29–36, 1976.
165. Birkbichler PJ, Patterson MK: Ann NY Acad Sci 312:354–365, 1978.
166. Couchman JR, Rees DA, Green MR, Smith CG: J Cell Biol 93:402–410, 1982.
167. Hynes RO: In Post G, Nicholson GL (eds): "Cell Surface Review." Vol 7. Amsterdam: Elsevier/North-Holland, pp 97–136, 1982.
168. Goldman RD, Pollard T, Rosenbaum J (eds): "Cell Motility." Cold Spring Harbor Press, 1976.
169. McNutt NW, Culp LA, Black PH: J Cell Biol 56:413–428, 1973.
170. Schollmeyer JV, Furcht LT, Goll DE, Robson RM, Stromer MH: Cold Spring Harbor Press, 1976.
171. Wehland J, Osborn M, Weber K: J Cell Sci 37:257–273, 1979.
172. Geiger B, Tokuyasu KT, Dutton AH, Singer SJ: Proc Natl Acad Sci USA 77:4127–4131, 1980.
173. Burridge K, Feramisco J: Cell 19:587–595, 1980.
174. McNutt NS, Culp LA, Black PH: J Cell Biol 50:691–708, 1971.

175. Goldman RD, Yerna MJ, Schloss JA: In Revel JP, Henning V, Fox CF, (eds): "Cell Shape and Surface Architecture." New York: Alan Liss, 1978.
176. Pollack R, Rifkin D: Cell 6:495–506, 1978.
177. Ash JF, Vogt PI, Singer SJ: Proc Natl Acad Sci USA 73:3603–3607, 1976.
178. Pouyssegur J, Pastan I: Exp Cell Res 121:373–382, 1979.
179. Willinsham MC, Yamada K, Yamada S, Pouyssegur J, Pastan I: Cell 10:375–380, 1977.
180. Ali IU, Hynes RO: Biochim Biophys Acta 471:16–21, 1977.
181. Furcht LT, Wendelschafer-Crabb G: Exp Cell Res 114:1–4, 1978.
182. Kurkinen M, Wartiovaara L, Vaheri A: Exp Cell Res 111:127–137, 1977.
183. Hynes RO, Destree AT: Cell 15:875–886, 1978.
184. Heggeness MH, Asit JF, Singer SJ: Ann NY Acad Sci 312:414–417, 1978.
185. Perdue JF: J Cell Biol 58L:265–283, 1973.
186. Culp LA, Bensusan H: Nature 273:683–682, 1978.
187. Ali IU, Hynes RO: Cell 14:439–446, 1978.
188. Postlethwaite AE, Seyer JM, Kang AH: Proc Natl Acad Sci USA 75:871–875, 1978.
189. Postlethwaite AE, Synderman R, Kang AH: J Exp Med 144:1188–1203, 1976.
190. Albrecht-Buehler G: Cell 395–404, 1977.
191. Akers RM, Mosher DF, Lilien JE: Dev Biol 86:179–188, 1981.
192. Orly J, Sato G: Cell 17:295–305, 1975.
193. Aberchrombi M: Nature 281:259–263, 1979.
194. Grinnell F, Billingham RE, Burgess L: J Invest Dermatol 76:181–189, 1981.
195. Repesch LA, Fitzgerald TJ, Furcht LT: J Histochem Cytochem 30:351–358, 1982.
196. Critchley DR, England MA, Wakely J, Hynes RO: Nature 280:498–500, 1979.
197. Mayer BW, Hay ED, Hynes RO: Dev Biol 82:267–286, 1981.
198. Mosher DF: J Biol Chem 251:1639–1645, 1976.
199. Gauss-Muller V, Kleinman HK, Martin GR, Schiffman E: J Lab Clin Med 96:1071–1080, 1980.
200. Zigmond SH: J Cell Biol 75:606–616, 1977.
201. Greenberg JH, Seppa S, Seppa H, Hewitt AT: Dev Biol 87:259–266, 1981.
202. Grotenhorst GR, Seppa HE, Kleinman HK, Martin GR: Proc Natl Acad Sci USA 78:3669–3672, 1981.
203. Bunge RP, Bunge MB: J Cell Biol 78:943–950, 1978.
204. Perri RT, Kay NE, McCarthy J, Vassella RL, Jacob HS, Furcht LT: Blood 60:430–435, 1982.
205. Yoshida K, Ozaki T, Ushijima K, Hayashi H: Int J Cancer 6:123–132, 1970.
206. Ozaki T, Yoshida K, Ushijima K, Hayashi H: Int J Cancer 7:93–100, 1971.
207. Romualdez AG Jr, Ward PA: Proc Natl Acad Sci USA 72:4128–4132, 1975.
208. Orr W, Phan SH, Vaheri A, Ward PA, Kreutzer DL, Webster RO, Henson PM: Proc Natl Acad Sci USA 76:1986–1989, 1979.
209. Orr FW, Varani J, Delikatny J, Jain N, Ward PA: Am J Pathol 102:160–167, 1981.
210. Kort EN, Goy MF, Larsen SH, Adler J: Proc Natl Acad Sci USA 72:3939, 1975.
211. O'Dea RF, Vivenos OH, Axelrod J, Aswanikumar S, Schiffmann E, Corcoran BA: Nature 272:462–464, 1978.
212. Pike MC, Kredich NM, Snyderman R: Proc Natl Acad Sci USA 75:3928, 1978.
213. Schiffmann E, Gallin JI: Curr Top Cell Regul 15:203–261, 1979.
214. Yaffe D, Dym H: Cold Spring Harbor Symp Quant Biol 37:543–571, 1972.
215. Podleski TR, Greenberg I, Schlessinger J, Yamada K: Exp Cell Res 122:317–326, 1979.
216. Woodbridge PA: PhD Thesis, University of Minnesota, 1981.
217. Puri EC, Chiquet M, Turner DC: Biochem Biophys Res Commun 90:883–889, 1979.
218. Whatley R, NG K-C, McMurray WC, Sanwal BD: Biochem Biophys Res Commun 70:180–185, 1976.

219. Pennypacker JP, Hassell JR, Yamada K, Pratt RM: Exp Cell Res 121:411–415, 1979.
220. Dessau W, Sasse J, Timpl R, Jilek F, Von der Mark: J Cell Biol 79:342–355, 1978.
221. Dessau W, Vertel BM, Von der Mark H, Von der Mark K: J Cell Biol 90:78–83, 1981.
222. Weiss RE, Reddi AH: J Cell Biol 88:630–636, 1981.
223. West CW, Lanza R, Rosenbloom J, Lowe M, Holtzer H, Avdalovic N: Cell 17:491–501, 1979.
224. Spiegel E, Burger M, Spiegel M: J Cell Biol 87:309–313, 1980.
225. Labat-Robert J, de Ceccatty MP, Robert L, Auger C, Lahias C, Garrone R: In "Glycoconjugates: Proceedings of The Fifth International Symposium." Schauer R, Boer P, Buddecke E, Kramer MF, Vliegenthart JFG, Wiegandt H (eds): Georg Thiene Publishers, 1979, pp 431–432.
226. Kurkinen M, Alitalo K, Vaheri A, Stenman S, Saen L: Dev Biol 69:598–600, 1979.
227. Newgreen D, Thiery JP: Cell Tissue Res 211:269–291, 1980.
228. Greenberg JH, Seppa S, Seppa H, Hewitt T: Dev Biol 87:259–266, 1981.
229. Sieberblum M, Sieber F, Yamada K: Exp Cell Res 133:285–295, 1981.
230. Waterman R, Balian G: Anat Rec 198:619–635, 1980.
231. Thesleff I, Barrach J, Foidart JM, Vaheri A, Pratt RM, Martin GR: Dev Biol 81:181–192, 1981.
232. Zetter BR, Martin GR: Proc Natl Acad Sci USA 75:2324–2328, 1978.
233. Wartiovaara J, Leivo I, Vaheri A: Dev Biol 69:247–257, 1979.
234. Hynes RO, Bye JM: Cell 3:113–130, 1975.
235. Hynes RO: Proc Natl Acad Sci USA 70:3170–3174, 1973.
236. Gahmberg CG, Hakomori SI: J Biol Chem 250:2447–2451, 1975.
237. Gahmberg CG, Kiehn D, Hakomori SI: Nature 248:213–215, 1974.
238. Robbins PW, Wickus GS, Branton PE, Gaffney G, Hirschberg CB, Fuchs P, Blumberg PM: Cold Spring Harbor Symp Quant Biol 39:1173–1180, 1974.
239. Hogg NM: Proc Natl Acad Sci USA 71:489–492, 1974.
240. Ruoslahti E, Vaheri A, Kuusela P, Linder E: Biochim Biophys Acta 233:352–358, 1973.
241. Ruoslahti E, Vaheri A: J Exp Med 141:497–501, 1975.
242. Ruoslahti E, Vaheri A: Nature 248:790–791, 1974.
243. Clarkson B, Baserga R (eds): "Control of Proliferation in Animal Cells." Cold Spring Harbor, New York: Cold Spring Harbor Laboratory Press, 1974.
244. Yamada K, Ohanian SH, Pastan I: Cell 8:241–245, 1976.
245. Pealstein E, Waterfield MD: Biochim Biophys Acta 362:1–12, 1974.
246. Stenman S, Wariovaara J, Vaheri A: J Cell Biol 74:453–467, 1977.
247. Blumberg PM, Robbins PW: Cell 6:137–147, 1975.
248. Teng NNH, Chen LB: Proc Natl Acad Sci USA 72:413–417, 1975.
249. Pearlstein E, Hynes RO, Faanks LM, Hemmings VJ: Cancer Res 36:1475–1480, 1976.
250. Keski-Oja J, Vaheri A, Ruoslahti E: Int J Cancer 17:261–269, 1976.
251. Blumberg PM, Driedger PE, Rossow PW: Nature 264:446–447, 1976.
252. Keski-Oja J, Vaheri A, Ruoslahti E: Int J Cancer 17:261–269, 1979.
253. Hynes RO, Destree A: Proc Natl Acad Sci USA 74:2855–2859, 1977.
254. Keski-Oja, Todaro G: Cancer Res 40:4722–4727, 1980.
255. Mosher DF, Saksela O, Keski-Oja J, Vaheri A: J Supramol Struct 6:551–557, 1977.
256. Adams SL, Sorbel ME, Howard BH, Olden K, Yamada K, de Crombruggh B, Pastan I: Proc Natl Acad Sci USA 74:3399–3403, 1977.
257. Wagner DP, Ivatt R, Destree AT, Hynes RO: J Biol Chem 256:11708–11715, 1981.
258. Furcht LT, Mosher DF, Wendelschafer-Crabb G, Woodbridge PA, Foidart JM: Nature 277:393–395, 1979.
259. Furcht LT, Mosher DF, Wendelschafer-Crabb G, Foidart JM: Cancer Res 39:2077–2083, 1979.

260. Smith D: University of Minnesota PhD thesis, 1982.
261. Marceau N, Goyette R, Valet P, Deschenes: J Exp Cell Res 125:497–502, 1980.
262. Hayman E, Engvall E, Ruoslahti E: Exp Cell Res 127:478–481, 1980.
263. Gerfaux J, Rousette F, Cahny-Fournier F, Chany C: Cancer Res 41:3629–3634, 1981.
264. Pfeffer L, Wang E, Tamm I: J Cell Biol 85:9–17, 1980.
265. Nielson SE, Puck TT: Proc Natl Acad Sci USA 77:985–989, 1979.
266. Chen LB, Gudor RD, Sun TT, Chen A, Mosesson MW: Science 197:776–778, 1977.
267. Mosher DF, Vaheri A: Exp Cell Res 112:323–334, 1978.
268. Yamada KM, Yamada SS, Pastan I: J Cell Biol 74:649–654, 1977.
269. Hata R, Peterkofsky B: Proc Natl Acad Sci USA 74:2933–2937, 1977.
270. Kraemer PM: Biochemistry 10:1445–1451, 1971.
271. Kraemer PM, Tobey RA: J Cell Biol 55:713–715, 1972.
272. Roblin R, Albert SO, Gleb NA, Black PH: Biochemistry 14:347–357, 1975.
273. Davidson EA, MacPherson I: Exp Cell Biol 95:218–222, 1975.
274. Culp LA: Biochemistry 15:4094–4104, 1976.
275. Chiarugi VP, Vannucchi S, Urbano P: Biochim Biophys Acta 345:283–293, 1974.
276. Burridge K: Proc Natl Acad Sci USA 73:4457–4461, 1976.
277. Itaya K, Hakomori SI: FEBS Lett 66:65–69, 1976.
278. Clarke SN, Fink LM: Biochim Biophys Acta 464:433–441, 1977.
279. Der CJ, Stanbridge EJ: Cell 15:1241–1251, 1978.
280. Kahn P, Shin SI: J Cell Biol 82:1–16, 1979.
281. Eun CK, Klinger HP: Cytogenet Cell Genet 27:57–64, 1980.
282. Smith M, Gold LI, Pearlstein E, Krinsky A: In "Human Gene Mapping: Fifth International Workshop on Human Gene Mapping." New York: Alan R. Liss, 1979.
283. Ruoslahti E, Klinger HP: Cytogenet Cell Genet 28:271–279, 1980.
284. Owerbach D, Doyle D, Shows TB: Proc Natl Acad Sci USA 75:5640–5644, 1978.
285. Rennard S, Church R, Rohrvach D, Shupp A, Abe S, Hewitt J, Murray J, Martin G: Biochem Genet 19:551–566, 1981.
286. Fagan JB, Yamada K, de Crombrugghe B, Pastan I: Nucl Acids Res 6:3471–3481, 1979.
287. Fagan JB, Sobel ME, Yamada K, de Crombrugghe, Pastan I: J Biol Chem 256:520–525, 1981.
288. Stathakis NE, Mosesson MW, Chen AB, Galanakis DK: Blood 51:1211–1222, 1978.
289. Stathakis N, Mosesson MW: J Clin Invest 60:855–865, 1977.
290. Mosher DF, Williams EM: J Lab Clin Med 91:729–735, 1978.
291. Ruoslahti E, Vaheri A: J Exp Med 141:497–501, 1975.
292. Mosher DF: J Biol Chem 251:1639–1645, 1976.
293. Mosher DF, Schad PE, Kleinman HK: J Clin Invest 64:781–787, 1979.
294. Mosher DF, Schad PE, Vann JM: J Biol Chem 255:1181–1188, 1980.
295. Mosher DF, Schad PE, Kleinman HK: J Supramol Struct 11:227–235, 1979.
296. Plow EF, Birdwell CR, Ginsberg MH: J Clin Invest 63:540–543, 1979.
297. Zucker MB, Mosesson MW, Broekman MJ, Kaplan K: Blood 54:8–12, 1979.
298. Ginsberg MH, Painier RG, Forsyth J, Birdwell CR, Plow EF: Proc Natl Acad Sci USA 77:1049–1053, 1980.
299. Hynes RO, Ali IU, Destree AT, Mautner V, Perkins ME, Senger DR, Wagner DD, Smith KK: Ann NY Acad Sci 312:317–342, 1978.
300. Muggli R, Baumgartner HR: Thrombosis Res 3:715, 1973.
301. Marguerie GA, Plow EF, Edgington TS: J Biol Chem 254:5357–5363, 1979.
302. Jaffee RM: In Gordon JL (ed): Amsterdam: Elsevier/North Holland, 1976, pp 262–292.
303. Stemerman MB, Spaet TH: Bull NY Acad Med 48:289–301, 1972.
304. Cazenave JP, Packham MA, Mustard JF: J Lab Clin Med 82:978, 1973.
305. Mason RG: Prog Hemo 1:141–164, 1972.

306. Baumgartner HR, Muggli R: In Gorden JL (ed): "Platelets in Biology and Pathology." Amsterdam: Elsevier/North-Holland, 1976.
307. Jaffe R, Deykin D: J Clin Invest 53:875, 1974.
308. Fauvel F, Legrand YJ, Kuhn K, Bentz H, Fietzek PP, Caen JB: Thromb Res 16:269–273, 1979.
309. Fauvel F, Legrand Y, Caen JP: Thromb Res 12:273–285, 1978.
310. Koteliansky VE, Leytin VL, Sviridov DD, Repin VS, Smirnov VN: FEBS Lett 123:59–62, 1981.
311. Stemerman MB, Baumgartner HR, Spaet TH: Am J Pathol 24:179–186, 1971.
312. Huang TW, Lagunoff D, Benditt EP: Lab Invest 31:156–160, 1975.
313. Huang TW, Benditt EP: Am J Pathol 92:99–110, 1978.
314. Balleisen L, Gay S, Marx R, Kuhn H: Klin Wschn 53:903–905, 1975.
315. Tschopp T, Weiss HJ, Baumgartner HR: J Lab Clin Med 83:296–300, 1974.
316. Sakariasen KS, Bolhvis PA, Sixma JJ: XVII Cong Int Soc Haematol 529, 1978.
317. Nyman D: Thromb Res 10:743–751, 1977.
318. Legrand YJ, Rodriguez-Zebauos A, Dartalis G, Fauvel F, Chen JP: Thromb Res 13:909–911, 1978.
319. Nurden AT, Caen JP: Br J Haematatol 28:253–260, 1974.
320. Saba TM: Arch Internal Med 126:1031, 1970.
321. Benacerraf B, Biozzi G, Halpern BN, Stiffel C: Springfield, Illinois: Charles Thomas, 1967, pp 52–79.
322. Saba TM, Jaffe E: Am J Med 68:577–594, 1980.
323. Mosher DF, Proctor RA, Grossman JE: Adv Inflammation Res 2:187–207, 1981.
324. Saba TM, di Luzio NR: Am J Physiol 216:197–205, 1969.
325. Filkins JP, Chase RE, Smith JJ: J Reticuloendothelial Soc 2:287, 1965.
326. Murray IM: Am J Physiol 204:655–654, 1963.
327. Saba TM, Filkins JP, di Luzio NR: J Reticuloendothelial Soc 3:398, 1966.
328. Laurell CB: Anal Biochem 15:45–52, 1966.
329. Stathakis NE, Fountas A, Tsianos D: J Clin Pathol 34:504–508, 1981.
330. Scovill WA, Annest SJ, Saba TM, Blumenstock FA, Newell JC, Stratton HH, Powers SR: Surgery 86:284–293, 1979.
331. Saba TM, Scovill WA, Powers SR: Surg Ann 12:1–20, 1980.
332. Scovill WA, Saba TM, Blumenstock FA, Bernard H, Powers SR Jr: Ann Surg 188:521–529, 1978.
333. Saba TM, Blumenstock FA, Scovill WA, Bernar H: Science 201:622–624, 1978.
334. Ehrlich MI, Krushell JS, Blumenstock FA, Kaplan JE: J Lab Clin Med 98:263–271, 1981.
335. Priest JR, Ramsay NKC, Bennett AJ, Krivit W, Edson JR: J Ped 100:990–995, 1982.
336. Czop J, Kadish J, Austen F: Proc Natl Acad Sci USA 78:3649–3653, 1981.
337. Kuusela P: Nature 275:718–720, 1978.
338. Mosher DF, Proctor RA: Science 290:927–292, 1980.
339. Verbrugh HA, Peterson PK, Smith D, Nguyen B, Hoidal J, Wildinson B, Verhoef J, Furcht LT: Infect Immun 33:811–819, 1981.
340. Rajbhandry VL, Baddiley J: Biochem J 87:429–435, 1963.
341. Schleifer KH: Z Immun-Forsch 149:104–117, 1975.
342. Forsgren A: Infect Immun 2:672–673, 1970.
343. Woods DE, Bass JA, Johansson WG, Straus DC: Infect Immun 30:694–699, 1980.
344. Woods DE, Straus DC, Johanson WG, Bass JA: J Infect Dis 143:784–790, 1981.
345. Balian G, Click EM, Crouch E, Davidson JM, Bornsten P: J Biol Chem 254:1429–1432, 1979.

346. Smith DE, Furcht LT: J Biol Chem 257:6518–6523, 1982.
347. Smith DE, Mosher DF, Johnson RB, Furcht LT: J Biol Chem 257:5831–5838, 1982.
348. Hormann H, Seidle M: Hoppe-Seyler's Z Physiol Chem 361:1449–1452, 1980.
349. Sekiguchi K, Hakomori SI: Proc Natl Acad Sci USA 77:2661–2665, 1980.
350. Petersen TE, Thorgersen HC, Skorstengarrd K, Vibe-Pedersen K, Sahl P, McDonagh R, McDonagh J, Magnusson S, Sottrup-Jense L: Thrombos Haemostasis 46:83, 1981.
351. Furie M, Frey A, Rifkin D: J Biol Chem 255:4391–4394, 1980.
352. Ruoslahti E, Hayman E, Engvall E: J Biol Chem 256:7277–7281, 1981.
353. Hahn LE, Yamada KM: Cell 18:1043–1051, 1979.
354. Hahn L-HE, Yamada K: Proc Natl Acad Sci USA 76:1160–1163, 1979.
355. McDonald JA, Kelley DG: J Biol Chem 255:8848–8858, 1980.
356. Keski-Oja J, Sen A, Todaro GJ: J Cell Biol 581:237–251, 1979.
357. Keski-Oja J, Yamada K: Biochemistry 193:615–620, 1981.
358. Keski-Oja J, Rouslahti E, Engvall E: Biochim Biophys Acta 100:1515–1522, 1981.
359. Zardi L, Siri A, Carnemolla B, et al: Cell 18:649–657, 1979.
360. Akiyama SK, Yamada K, Hayashi M: J Supramol Struct Cell Biochem 16:345–358, 1981.
361. Richter H, Seidl M, Hormann H: Hoppe-Seyler Z Physiol Chem 362:399–408, 1981.
362. Yamada K, Kennedy DW, Kimata K, Pratt RM: J Biol Chem 255:6055–6063, 1980.
363. Hayashi M, Schlesinger DH, Kennedy DW, Yamada K: J Biol Chem 255:10017–10020, 1980.
364. Hayashi M, Yamada K: J Biol Chem 257:5263–5267, 1982.
365. Sekiguchi K, Fukuda M, Hakomori S: J Biol Chem 256:6452–6462, 1981.
366. Laterra J, Culp L: J Biol Chem 257(2):719–726, 1982.
367. Pierschbacher MD, Hayman EG, Ruoslahti E: Cell 26:259–267, 1981.
368. Sekiguchi K, Hakomori S: Biochem Biophys Res Commun 97:709–715, 1980.
369. Fukuda M, Hakomori S: J Biol Chem 254:5442–5450, 1979.
370. Hynes RO, Destree A: Proc Natl Acad Sci USA 74:2855–2859, 1977.
371. McConnell MR, Blumberg PM, Rossow PW: J Biol Chem 253:7522–7530, 1978.
372. Ali IU, Hynes RO: J Biol Chem 253:7522–7530, 1978.
373. Fyrand O, Munthe E, Solum HO: Ann Rheum Dis 37:347–350, 1978.
374. Stenman S, Von Smitten K, Vaheri A: Acta Med Scand 642:165–170, 1980.
375. Clark RAF, Dvorak HF, Colvin RB: J Immunol 126:787–793, 1981.
376. Resnick DG, Nachman LP: Blood 57:339–342, 1981.
377. Weiss MA, Doi BS, Ooi YM, Engvall E, Ruoslahti E: Lab Invest 41:340–347, 1979.
378. Linder EA, Miettinen, Tornroth T: Lab Invest 42:70–75, 1980.
379. Rennard SI, Crystal RG: JCI 69:113–122, 1981.
380. Villiger B, Broekelmann T, Kelley D, Heymach G, McDonald JA: Am Rev Respir Dis 124:652–654, 1981.
381. Cooper SM, Keyser AJ, Beaulien AD, Ruoslahti E, Nimmni MF, Quismorio FP: Arthritis Rheum 22:983–987, 1979.
382. Soria J, Soria C, Ryckewaert JJ, Naveau B, Lafay P, Ryckewaert A: Arthritis Rheum 23:1334–1335, 1980.
383. Scott DL, Farr M, Crockson AP, Walton KW: Clin Sci 62:71–76, 1982.
384. Chandrasekhar S, Millis AJT: J Cell Phys 103:47–54, 1980.
385. Saba TM, Cho E: Adv Shock Res 3:251–271, 1980.
386. Postlethwaite AE, Keski-Oja J, Balian G, Kang AH: J Exp Med 153:494–499, 1981.

Modern Cell Biology, 1:119–170

Membrane Biogenesis, Enveloped RNA Viruses, and Epithelial Polarity

Enrique Rodriguez-Boulan

From the Department of Pathology, State University of New York, Downstate
Medical Center, Brooklyn, New York 11203

I. INTRODUCTION

The subject of epithelial polarity has received considerable attention from the physiologists for more than three decades. The attraction was derived from the fascinating ability of epithelial membranes to generate steep concentration gradients of nutrients and electrolytes between diverse external media and the *internal milieu* of the body. This ability has a molecular basis in the asymmetric distribution of enzymes and transporting systems between the two plasmalemmal domains of the epithelial cell: apical, which relates to the external medium, and basolateral, continuous with the *internal milieu* and the blood and therefore subject to modulation by hormones and other regulatory factors. The impressive advance in recent years in our understanding of the biogenesis of plasma membrane proteins opens the way for the exploration of epithelial polarity with a cell biological approach. From both theoretical and methodological standpoints, the field is ready to allow the study of the structural "signals" for polarity in the plasma membrane proteins and the systems that the cell utilizes to decode that information and mediate the delivery and retention of the proteins at the correct surface domain.

In the first part of this paper, a bird's-eye view of three research fields related to plasma membrane biogenesis is presented: secretion, glycoproteins of enveloped RNA viruses, and membrane recycling. The second part of the review deals more specifically with recent advances in the epithelial field, with particular emphasis on a model system developed in our laboratory which uses the glycoproteins of enveloped RNA viruses to study the mechanisms of polarization of plasma membrane proteins in epithelial cells.

II. BIOGENESIS OF CELLULAR MEMBRANES:
A. Biosynthesis and Distribution of Organellar Components

Throughout its life a cell has to deal with the problem of generating and keeping as separate entities a diverse array of organellar membranes and the compartments they circumscribe. Each of these membranes and compartments is endowed with a well-defined set of molecular components: lipids, glycolipids, proteins, glycoproteins, nucleic acids. Furthermore, membrane components have a characteristically asymmetric distribution with respect to the plane of the bilayer. The production of molecules destined to these diverse subcellular membranes and compartments is carried out, however, in a few biosynthetic sites (Table I)—two each for proteins, lipids, glycoproteins, and nucleic acids (in eukaryotic cells). Specific distribution systems must exist, therefore, to ensure that each organellar membrane or compartment receives only its own components, adequately processed, so that it can perform its programmed function. Only rarely, if ever, are cell membranes or compartments generated by "de novo" assembly of their components. (One possible exception is given by the membranes of some pox viruses, which are assembled "de novo" in the cytoplasm [32]; the nuclear membrane, which disintegrates during cell division, is regenerated from internal elements of the endoplasmic reticulum (ER) after cytokinesis.) Organellar growth is achieved by the incorporation of new material into a preexisting set of organelles inherited during cell division.

The energy sources, biosynthetic precursors, and enzymes necessary for the synthesis of most organellar proteins and lipids are localized in the cytosol or associated with the cytoplasmic side of a specialized membrane, the endoplasmic reticulum (ER) (as a partial exception, mitochondria and chloroplasts are able to synthesize some of their own proteins and lipids). Likewise, the enzymes for the synthesis of the dolichol-phosphate-linked oligosaccharide precursor of glycoproteins appear to be also localized on the cytoplasmic side of the ER membrane [17]. In order to reach their final localization and specific orientation in organellar membranes or spaces, proteins and lipids must utilize special transporting systems which assist them in traversing the hydrophobic membrane barriers, and in migrating from one cell organelle to another. It is believed that specific structural markers or signals in proteins operate as "zip codes," informing the transporting systems of the particular address where the polypeptide is to be delivered. According to this view, proteins sharing a common destination must possess similar "zip codes." Some of these signals have been identified as transient extra peptides found in the precursor state of the proteins (or "pre-proteins") such as the hydrophobic, amino terminal sequence of 15–30 amino acids found in precursors of secretory and integral transmembrane proteins and the extra peptides of cytosolic precursors of some mitochondrial and chloroplast proteins [6, 20–23]. Other distribution signals may be permanent features of the polypeptide chain, such as those which direct the segregation of ovalbumin into the ER lumen [24], the insertion of cytochrome P450 and erythrocyte band 3 into the ER membrane [25, 26, 26a] or the delivery from cytosol to peroxisomes of catalase and uricase [27]. Finally, signals may be added co- or posttranslationally to the proteins as, for example, the mannose-6-phosphate residues which direct lysosomal proteins from their site of synthesis,

TABLE I. Sites of Biosynthesis of Organellar Components

Biosynthetic site	Destination	Examples	References
Proteins			
Cytosolic polysomes			
Free	Cytoplasmic matrix	Soluble, ribosomal, and cytoskeletal proteins	1, 2
	Nuclear envelope, ER, Golgi, plasma membrane	Peripheral and integral (nontransmembrane) proteins on cytoplasmic face	2, 3, 4
	Mitochondria	Most proteins in IMM, several proteins in mitochondrial matrix, intermembrane space and OMM	2, 4, 5, 6
	Chloroplasts	Most proteins in thylakoid membranes; several proteins in the other chloroplast membranes and compartments	6
	Peroxisomes	Luminal proteins (catalase, peroxidase)	2, 4, 5, 11
	Nuclear matrix	Histone and nonhistone proteins	16
Bound to ER and Nuclear membrane	Nuclear membrane, ER, Golgi, PM	Integral (transmembrane and nontransmembrane), peripheral proteins on the cisternal (external) face	2, 3, 4, 5
	Secretion	Secretory proteins	1, 7, 8
	Lysosomes	Luminal and membrane (?) proteins	9, 10
	Peroxisomes	Membrane proteins(?)	10, 11

Mitochondrial and chloroplast polysomes	Mitochondria chloroplasts	Integral membrane and soluble proteins	6
Lipids			
ER	Most cellular membranes and secretion	Most phospholipids, cholesterol, and dolichol P	12
Mitochondria	Mitochondria	Cardiolipin, phosphatidic acid	12
Glycoproteins			
ER	ER, Golgi, lysosomes, PM, secretion	Adition of glycoprotein core (GLc NAc_2 Man_9 Glc_3) from dolichol-P to luminal and integral membrane proteins	13, 14
Golgi	Golgi, lysosomes, PM, secretion	"trimming" of glycoprotein side chains, and addition of terminal trisaccharides (GlcNac-Gal-SA)	
Nucleic acids			15
Nucleus	Nucleus	DNA	
	Cytoplasm	tRNA, mRNA, rRNA	
Mitochondria	Mitochondria	DNA, tRNA, mRNA, rRNA	

the endoplasmic reticulum, to the lysosomal compartment [9, 28]. Very little detailed knowledge is available, however, on the way these signals operate or on the cellular recognition mechanisms which decipher them during the sorting process.

Even less well understood are the distribution mechanisms by which different cell membranes acquire their characteristic complement of lipids and how these lipids become asymmetrically distributed with respect to the plane of the bilayer. Thus, any model for organelle biogenesis must provide answers to the following key cell biological questions:

1. Where are the different organellar components synthesized?
2. What pathways are used for their distribution to their site of function?
3. What are the sorting signals involved in this distribution? Are they transient or permanent? In the cases of proteins, are they added co- or posttranslationally? How many signals participate in the distribution of a particular component?
4. What are the cellular mechanisms that "read" and "decode" the sorting signals?

In conclusion, specific distribution signals and mechanisms are responsible for the delivery of molecules from relatively few sites of synthesis to many different destinations in the cell. Some of these signals have been identified but most are still unknown. Even for those that have been recognized (peptides in presecretory proteins, premitochondrial and prechloroplast proteins and mannose-6-phosphate residues of lysosomal proteins), their mode of operation is not yet understood. Several recent reviews have extensively discussed the cellular distribution mechanisms for proteins [2, 4–6, 29–31]. The origins of protein and lipid asymmetry in biological membranes have also been reviewed [3, 32, 33]. In this review, we deal with the problem of distribution of plasma membrane proteins, in particular those of epithelial cells. Because plasma membrane proteins follow, during their biogenesis, an intracellular pathway that parallels that of secretory proteins, the current concepts on the secretory pathway will be briefly reviewed.

B. The Secretory Pathway

The classical studies of Palade and his collaborators on enzyme secretion by exocrine pancreatic cells provide the first detailed example of a protein distribution system operating in a eukaryotic cell [for reviews see 1, 7, 8, 34]. Although the original scheme was developed for glandular epithelial cells, it later became apparent that very similar mechanisms were utilized for secretion of macromolecules by every kind of eukaryotic cell, including macrophages, granulocytes, plasma cells, gametes, nerve cells, smooth muscle cells, and fibroblasts [1]. Production of polypeptides, polysaccharides, or glycoproteins for export appears thus to be a universal feature of all cells and, modified according to the cell type, the secretory pathway provides a conceptual framework to understand the mechanisms of synthesis, intracellular transport and exocytosis of macromolecules by eukaryotic cells.

The following are well-established steps in the synthesis of secretory proteins by a pancreatic cell [1]:

1. Synthesis of a precursor (or pre-protein) by polysomes attached to the outer surface of the endoplasmic reticulum (ER) [35–37]. The attachment of polysomes to the ER membrane is mediated by an amino terminal extra sequence of 15–25 amino acids of a predominantly hydrophobic nature, the "signal sequence" that interacts with a putative receptor in the ER membrane [38, 39]. An ionic interaction between the large ribosomal subunit and a receptor in the ER membrane provides extra strength to this bond [40–44].

2. Segregation into the ER cisternae. The conceptual framework for this process was originally coined by Blobel and Sabatini [45] as the "signal hypothesis" and later further elaborated by Blobel and Dobberstein [38, 39]. Through interaction with an ER membrane receptor, the signal sequence facilitates the cotranslational translocation of the polypeptide across this membrane, the energy for the process being that involved in the formation of the peptide bond. The transient existence of the extra peptide in presecretory proteins was demonstrated by their increased molecular weight (when compared to the secreted product or the pro-protein in the lumen of the ER) in cell-free translation experiments with mRNA coding for secretory polypeptides [18, 39, 40, 46–49]. The signal is cleaved cotranslationally by a signal peptidase on the cisternal side of the ER membrane. Additional cotranslational modifications include the proximal glycosylation of proteins to become glycoproteins, formation of disulfide bonds, and the hydroxylation of prolyl and lysyl residues in the case of collagen [50, 51].

3. Intracellular transport. After segregation in the lumen of the rough ER, pancreatic secretory proteins are transferred to transitional smooth ER vesicles and then to the Golgi complex in an energy-dependent step.

4. Concentration, storage, exocytosis. From the Golgi apparatus the proteins are transferred to condensing vacuoles where they are concentrated, without requirement of energy, and later stored in secretion granules for various lengths of time until, in an energy- and Ca^{++}-dependent step, they fuse with the apical plasmalemma for discharge into the acinar lumen. Because secretion in these cells is subject to regulation by external stimuli, it has been called "regulated secretion" [52].

In other secretory cells, such as fibroblasts, chondrocytes, and plasma cells, the concentration and storage steps are absent and discharge appears to take place continuously ("nonregulated" secretion [52]).

It has been suggested [1] that the secretory pathway might be involved in the distribution of membrane components destined to the different organelles in this pathway, including the plasma membrane. In order to demonstrate this hypothesis, it was necessary to follow a plasma membrane protein from the presumed site of synthesis—the ER—to the cell surface. Because the turnover rates of membrane proteins (and therefore their concentrations in intracellular precursor pools) are much lower than those of secretory proteins, the first confirmation of this hypothesis came from the study of model plasma membrane proteins that are *not* normally components of the cell: the membrane glycoproteins of enveloped RNA viruses.

C. Model Plasma Membrane Proteins: The Glycoproteins of Enveloped RNA Viruses

The best-studied viruses of this group fall into five major classes: rhabdo-, myxo-, paramyxo-, toga-, and retroviridae. Typically, these viruses possess a single-stranded RNA genome, segmented in the case of myxoviruses, which codes for a maximum of eight or nine proteins, including one to three envelope glycoproteins, one or more nucleoproteins, and RNA-dependent RNA polymerases (rhabdo-, myxo-, toga-, paramyxoviruses) or RNA-dependent DNA polymerases (retroviruses). Progeny virions usually mature by budding from the plasmalemma of the infected cell as a result of a process that involves the recognition of patches of viral glycoproteins in the cell surface by viral nucleocapsids synthesized in the cytosol. A matrix (M) protein, synthesized by free polysomes and present on the cytoplasmic aspect of the bilayer in rhabdo-, myxo-, and paramyxo viruses, appears to participate in this recognition. Cellular surface proteins are largely excluded from the viral bud by poorly understood mechanisms, so that the viral envelope contains almost exclusively virally coded proteins. The viral lipids, however, appear to represent quite faithfully the composition of the host plasma membrane [32, 33]. The experimental advantages provided by these viruses for studies on plasma membrane biogenesis have been stressed in several reviews [4, 32, 33]. They are summarized briefly in the following points:

- The limited genome of these viruses, encoding only a few proteins, ensures that they will use mostly cellular mechanisms for their replication, which turns them into useful probes to study those mechanisms.
- Infection by several of these viruses abolishes cellular protein synthesis. Thus, the cell manufactures only virally coded proteins, one or two of them being integral plasma membrane proteins. This extremely simplified situation, in which plasma membrane biogenesis accounts for 20–30% of the cellular biosynthetic activity, represents a great experimental advantage over the situation in the uninfected cell, where the synthesis of plasma membrane components is a very small percentage of the cellular protein and lipid synthesis.
- The viral glycoproteins are easily purified, and sequence information is already available for several of them [53–56]. Cloning of their mRNAs is easier than

for cellular proteins, and temperature-sensitive mutants in which the migration
to the cell surface is impaired are available or can be generated with relative
facility.

Extensive studies carried out on the biogenesis of viral envelope glycoproteins in recent years have elegantly confirmed the prediction that plasma
membrane proteins utilize for their biogenesis, to a large extent, the organelles
in the secretory pathway [4, 33, 55, 57]. They are synthesized by polysomes
bound to rough ER membranes [58, 59] via N-terminal sequences which
result in insertion of the proteins into the RER membrane. Differently from
secretory proteins, the translocation is not complete and the viral glycoproteins remain anchored to the ER membrane by hydrophobic sequences [55,
60, 61]. They are glycosylated cotranslationally via oligosaccharide lipid
donors and subsequently migrate from rough to smooth membranes during
which time their oligosaccharide moieties are processed by cellular enzymes
[4, 13, 14, 63–64]. The viral glycoproteins later migrate to the Golgi apparatus, as deduced from the addition of terminal sugars (N-acetylglucosamine, galactose, and sialic acid) by glycosyl transferases typical of this
organelle [14], after which the glycoproteins make their appearance in the
plasma membrane. This pathway has been recently demonstrated by immunoelectron microscopy for the glycoprotein (G) of vesicular stomatitis
virus (VSV, a rhabdovirus) [65] for the glycoproteins of semliki forest virus
(SFV, a togavirus) [65a] and for the hemagglutinin of influenza virus [227].
Recent work suggests that clathrin-coated vesicles transport mature G protein
at two different intracellular stages [66, 67]. Additional processing for some
envelope proteins may involve proteolytic cleavage, desialylation by viral
neuraminidases, sulfation, and covalent attachment of fatty acids [32, 33].
The findings for viral glycoproteins have recently been extended to several
cellular integral plasma membrane proteins such as H_2 and HLA [72, 73]
antigens and erythrocyte band 3 [26, 26a].

In spite of the increase in our general knowledge on the intracellular
pathway and biochemical processing of plasma membrane proteins, several
important aspects such as how the translocation across the ER membrane
operates and what cellular structures are involved in their delivery from ER
to Golgi and from Golgi to plasma membrane, remain the subject of con-
troversy. Moreover, the key question of what is the information in membrane
proteins (sorting out signals) which determines that some of them remain in
the ER after synthesis, while others move ahead to Golgi, lysosomes, peroxisomes, or plasma membrane is still unanswered.

D. Endocytosis, Membrane Recycling, and the Golgi Complex

Evidence accumulated during recent years has demonstrated the existence
of a complex intracellular traffic of vesicles, mediating the transport of macromolecules between the cell surface and intracellular compartments and

between different cell compartments. This traffic is organized into major pathways or "roadways" along which successive membrane bound compartments become intermittently continuous by specific membrane fusion/fission events. Typically, membrane or luminal material from the proximal compartment is incorporated into a budding vesicle which later fuses with a distal compartment. Although there is considerable evidence in some cell systems that selectivity in the transfer of material operates at both the proximal and distal ends of the vesicular trajectory, there is no information on the nature of the mechanisms responsible for it. A related, unanswered question is how the various organellar membranes undergoing such extensive fusion events manage to retain their structural and biochemical individuality in spite of their well-demonstrated fluid nature [74]. It is possible that characteristic integral membrane proteins may exist in each organelle, such as the ribophorins in the rough ER [75], which bear ribosomal binding sites, and the recently described Golgi-specific proteins [76, 76a], that could provide a specific scaffolding recognized by other proteins destined to that membrane.

The Golgi complex (Fig. 1) is located at the crossroads of this vesicular traffic and plays an important role in its regulation [for recent reviews on this subject, see 77–79]. Its structure has been extensively reviewed [80–82]. It is usually placed in a juxtanuclear position and consists of a characteristic stack of 3–15 cisternae, closely apposed to each other, with associated vesicles and vacuoles. The most proximal cisterna, facing the nucleus, is usually called "cis," "forming," or "convex" face of the Golgi, and usually relates to a set of proximal vesicles that often appear to be budding from transitional elements of the RER. The distal face (also called "trans," "mature," or "concave") is associated with the production of vesicles which in regulated secretory cells (see section II,C) are large condensing vacuoles and secretory granules, while in nonregulated secretory cells are smaller vesicles frequently of the coated type (see Fig. 1). The distal face also relates to an anastomosing network of acid phosphatase-containing tubules known as Golgi-endoplasmic reticulum lysosome (GERL) presumably involved in the formation of lysosomes. Proximal and distal Golgi cisternae exhibit differences in terms of osmium staining (higher in proximal cisternae), enzyme content (thiamine pyrophosphatase is concentrated in distal cisternae), and membrane thickness (higher in distal cisternae) [77–79].

The major pathways of vesicular flow, with particular emphasis in epithelial cells, are represented in Figure 1. They can be classified in 1) centrifugal (from intracellular compartments to the cell surface), 2) centripetal (from cell surface to intracellular compartments), and 3) transcellular (from apical to basolateral surface or vice versa).

1. Centrifugal route. The centrifugal route includes the migration of products synthesized by the RER such as secretory proteins [1, 7, 8], lysosomal

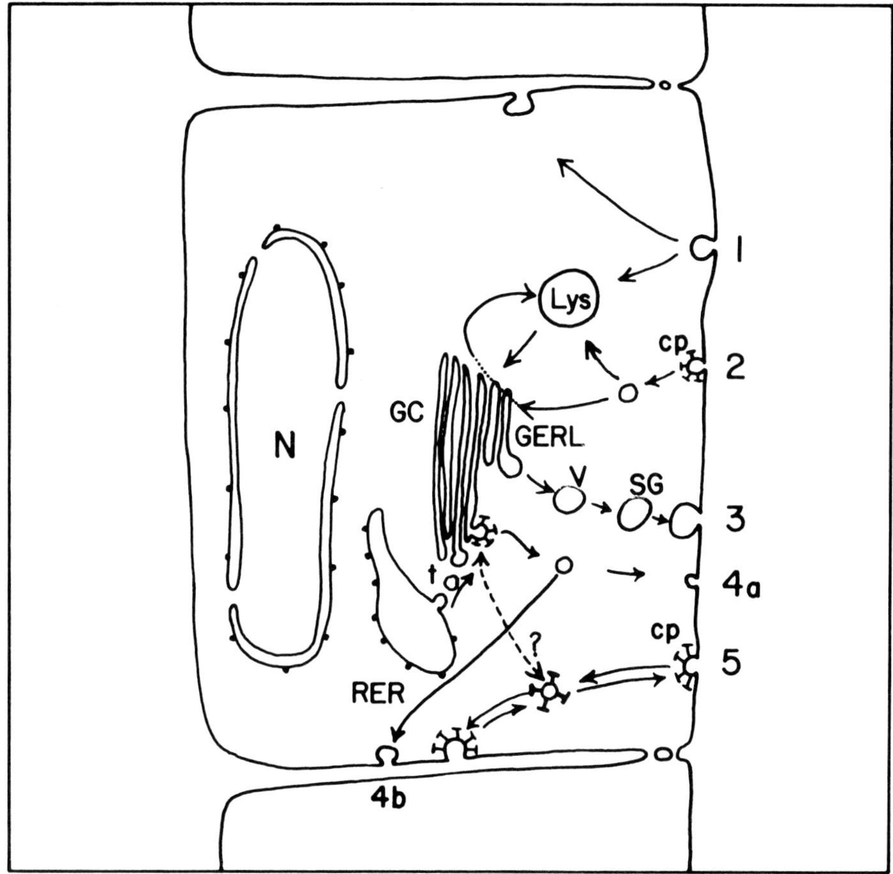

Fig. 1. Pathways of membrane recycling. This scheme represents several major pathways of exo- and endocytosis with special emphasis in epithelial cells. 1) Fluid-phase pinocytosis: material entering through this pathway is usually delivered to the lysosomes [94]. In endothelial cells, it constitutes a major mechanism for transcellular transport [110]. 2) Receptor-mediated endocytosis. Ligand-receptor complexes are clustered into coated pits (cp) and transported to lysosomes [99] or to the Golgi apparatus and later to the lysosomes [100, 101]. 3) Classical exocytotic pathway in "regulated" secretory cells [1, 52]. Proteins produced by the RER are delivered to the Golgi complex by transitional vesicles (t), or to condensing vacuoles (v), which may form from the GERL region [10], and become concentrated in secretory granules (SG) before release after fusion with the plasma membrane [1, 7, 8]. 4) Secretory pathway in "nonregulated" cells [1, 52] involves coated vesicles and small post-Golgi vesicles. Different populations of these vesicles may fuse specifically with the apical (4a) or the basolateral membrane (4b). This mechanism may be used for specific delivery of proteins to the apical or basal surface. 5) In several epithelial cell types, immunoglobulins are transported across the cell by a mechanism involving receptor-mediated endocytosis and coated vesicles, which largely bypasses the lysosomal compartment [103–106].

proteins [84], and membrane proteins for the various compartments in the secretory pathway, including the plasma membrane. The kinetics for the intracellular transport of these different classes of proteins, as determined by radioautographic studies or pulse-chase analysis of subcellular fractions, are rather similar. Secretory proteins reach the Golgi complex in 20–30 min and the extracellular medium in 60–90 min. Viral envelope glycoproteins [4, 57, 59], intrinsic plasma membrane proteins such as HLA [73] and H_2 antigens [72], and two liver surface glycoproteins [85] show similar kinetics (they reach the Golgi in approximately 20–30 min and the cell surface in about 60–80 min). Lysosomal proteins exit from the RER in about 15 min and arrive into lysosomes with a half-time of about 70 min [86]. A recent study comparing the migration of VSV G protein and several secretory proteins in a hepatoma cell line has demonstrated similar half-times for arrival to the cell surface [87].

The rapid migration of secretory and membrane proteins contrasts with the very low turnover rates measured for various membrane proteins [88–91], which display half-lives ranging from several hours to several days. In several exocrine cell types, it has been shown that the half-lives of membrane proteins in secretory granules are at least one order of magnitude longer than those of the secretory products contained in the granule [34, 92–93]. These results indicate that membrane proteins in the different segments of the secretory pathway, although migrating with similar kinetics as secretory proteins, are reutilized in several exocytotic cycles. Direct evidence for membrane recycling has been obtained using morphological techniques (see below).

2. Centripetal route. This pathway is easier to study with morphological techniques using tracers for fluorescence or electron microscopy. The processes by which portions of the surface membrane are interiorized as vesicles and vacuoles are generically known as endocytosis. It is currently accepted that endocytosis, like secretion, is a constitutive property of all eukaryotic cells [94]. The rate of endocytosis varies widely between different cell types, and may reach very high values. Macrophages and fibroblasts, for example, are able each hour to interiorize 186% and 54% of their surface area, respectively, as pinocytic vesicles, without alteration in cell volume and area [95a].

Two major types of endocytosis, fluid phase pinocytosis (FPP) and absorptive (receptor mediated) endocytosis, have been described (Fig. 1, pathways 1 and 2). Recent studies on the uptake of semliki forest virus by BHK cells suggest that the pathways for both may be the same [95]. Using tracers for the first one, such as peroxidase and dextrans, it has been shown that, in most cell types, the tracer is delivered to the lysosomes in a few minutes [94]. In some exocrine or endocrine cells, however, the tracers reach Golgi cisternae and secretory granules as well as lysosomes [96–98]. Studies following the fate of cationized ferritin, which binds electrostatically to the cell surface, have demonstrated that the marker can reach Golgi cisternae, GERL, condensing vacuoles and lysosomes [76, 97].

In receptor-mediated endocytosis, the binding of a ligand to a specific receptor enormously increases the efficiency of its interiorization. Receptor-ligand complexes are rapidly clustered into coated pits, which pinch off from the plasma membrane as coated vesicles. Low-density lipoprotein receptors interiorized via this mechanism by fibroblasts fuse within minutes with lysosomes [99]. Willingham and Pastan [100, 101] have recently postulated that, in fibroblasts, certain ligand-receptor complexes cluster into coated pits which, they believe, are permanently associated with the plasma membrane, and are later transferred into a special class of smooth vesicles, which they call "receptosomes." These vesicles display a longer existence and a saltatory movement in the cytoplasm, and, after several hours, they finally fuse with a region in the Golgi complex area, possibly the GERL [100, 101].

3. Transcellular route. Receptor-mediated endocytosis via coated pits and vesicles has been demonstrated for the transport of immunoglobulins across the epithelia of mucous membranes lining the digestive, respiratory, and urogenital tracts, as well as exocrine glands [102–104]. The vesicles appear to largely avoid the lysosomal compartment; instead, they traverse the cell and fuse with the opposite plasma membrane surface.

Polymeric immunoglobulins of the IgA or the IgM class, produced locally by plasma cells in the lamina propria, form a complex with a membrane-bound specific receptor in the epithelial basolateral surface, the *secretory component*, that is transported via endocytic coated vesicles to the apical surface (Fig. 1). Here, the complex is released into the luminal fluid, presumably after cleavage of the secretory component by luminal proteases [103, 105]. Transepithelial, apical to basal, transport of monomeric IgG molecules, responsible for the passive transfer of immunity from mother to young in mammals, is carried out, according to the species, by the digestive tract, placenta, or yolk sac and, in a reverse direction, by mammary gland cells [106]. The IgG molecules bind to apical receptors, specific for the Fc region, in a reaction favored by acidic pH (6–6.5), are interiorized by adsorptive endocytosis, and transported in coated vesicles across the cell to the basolateral surface [107–109], where the neutral pH of the internal medium promotes the dissociation of the complex.

Transepithelial transport of material incorporated via fluid-phase pinocytosis, very important in endothelial cells [110], does not appear to be a major pathway in epithelia, although it has been described for several epithelial membranes, such as kidney, intestine, and choroid plexus [111, 112].

Most of the material incorporated via this route appears to be delivered to lysosomes. In a recent paper, Abrahamson and Rodewald [113] demonstrated sorting of IgG and horseradish peroxidase into different vesicles after addition of both tracers to the intestinal lumen of newborn rats. Whereas IgG was incorporated into coated vesicles that bypassed the lysosomes and fused with the basolateral membrane, HRP was almost completely delivered to the lysosomal compartment, which suggests different pathways for receptor-mediated and fluid-phase endocytosis.

The evidence discussed above indicates the existence of mechanisms to interiorize specific receptors from the cell surface or to transport them between opposite domains of the epithelial plasmalemma. It is conceivable that similar processes may be involved in the generation and maintenance of the asymmetric distribution of surface components in epithelial cells (see section III, G)

III. EPITHELIAL CELLS AND THE SORTING OF PLASMA MEMBRANE PROTEINS

A. Polarity of Epithelial Cells

Epithelial cells may be classified into three main classes according to their prevalent function: covering, secretory, or absorptive. The two latter types of epithelial cells are endowed with a striking structural and functional polarity, responsible for their vectorial functions. The polarity is apparent both in their intracellular and surface organizations [114].

1. Intracellular polarity. Epithelial cells frequently display a typically asymmetric distribution of cell organelles and cytoskeletal elements in the cytoplasm. The basal one half or one third of the cell usually contains the nucleus, with the Golgi complex immediately above it. The mitochondria are concentrated in the regions with the highest requirements for ATP, such as close to the basolateral plasma membrane of intestinal and kidney tubular cells (rich in Na^+K^+ATPase, necessary for transepithelial transport of sodium), basal areas of secretory cells (rich in RER engaged in the synthesis of proteins for export), or the apical cytoplasmic regions of epithelial cells with high levels of pumping activity in the apical surface (choroid plexus, gallbladder). The free plasma membrane regions of epithelial cells are usually associated with high pinocytic and secretory activity, and therefore the cytoplasm contains variable amounts of vesicles and vacuoles, secretory granules, and lysosomes [114].

Epithelial cells possess the three types of cytoskeletal elements described in nonmuscle cells: actin-rich filaments, intermediate (10 nm) filaments, and microtubules. Actin filaments are usually found in association with all regions of the plasmalemma, forming a cortical meshwork in the basolateral surface

[150] but adopting a very regular organization in the apical brush border of absorptive epithelia.

The brush border of intestinal cells has been intensively studied in recent years as a model system for the organization of cytoskeletal structures in nonmuscle cells. A small intestine cell has approximately 1,000 microvilli, 0.1 to 0.15 μm in width by 1–2 μm in height [115, 116]. The core of the microvillus is formed by about 20–30 actin filaments, attached to the membrane at the microvillar tip and at multiple sites along the microvillus, via shorter filaments of uniform size (2 $\times$ 30 nm), regularly spaced every 6 nm [116–118]. Upon decoration with heavy meromyosin, the actin microfilaments exhibit uniform polarity, with the arrowheads running away from the point of membrane insertion [117, 118].

The microfilamentous bundles end proximally in a fibrillar meshwork located in a plane immediately under the apical surface, the *terminal web* [115, 116, 119]. In addition to being the zone of insertion of the microvillar filaments, the terminal web displays three distinct strata which are closely associated with the three components of the apical junctional complexes between intestinal cells: tight junctions, zonulae adherentes, and spot desmosomes:

- an *apical zone*, formed by a meshwork of thin microfilaments, which surrounds the lower, extramicrovillar one third of the actin filament bundles, and ends laterally in the cytoplasmic side of the tight junctional region of the lateral membrane;
- the intermediate *adherens zone,* a dense layer composed of interwoven thin and 10-nm filaments (tonofilaments) that inserts laterally on the second junctional area, the zonula adhaerens or belt desmosomes; and
- the *basal layer* formed by bundles of tonofilaments that end on the various spot desmosomes (maculae occludentes) distributed in the lateral surface [119].

Immunocytological and biochemical studies have demonstrated the presence of beta and gamma actin [120, 121], as in other nonmuscle cells, and two polypeptides, villin [122] and fimbrin [123], in the microvillar core; the terminal web was shown to be rich in myosin, tropomyosin, filamin, and alpha-actinin [121, 124–126]. Close to the point of its insertion in the zonula adhaerens, the terminal web exhibits an increased concentration of alpha-actinin, which is excluded within a 100–200 Å range of the membrane at the expense of another polypeptide, vinculin, which presumably plays a role in the attachment to the cell surface [125–127]. A similar relative distribution of these two proteins has been described in the attachment plaques of fibroblasts, the points of insertion of the actin stress fibers, and the dense plaques of smooth muscle cells [126–129]. Intermediate filaments usually run as bundles across the cytoplasm of the cell, interconnecting the spot desmosomes and thus contributing to determine the shape of the cell [116]. Each epithelial cell type appears to have a characteristic type of intermediate filament proteins, or *desmins,* as determined by bidimensional gel electrophoretic analysis [130].

Finally, microtubules are usually concentrated in the apical and peri-Golgi regions of intestinal cells [166] and are frequently found attaching to the apical membrane [132, 166]. The participation of microtubules in various cellular processes related to the apical surface has been reported. The evidence is usually derived from the action of microtubule depolymerizing drugs. For example, an inhibitory effect of colchicine on exocytosis in several secretory systems has been described [1, 131]. In toad and frog bladder epithelium, colchicine inhibits the fusion of particle-rich intracellular tubular structures with the apical plasma membrane. These particles, which are usually found associated with the E (exoplasmic) membrane face upon freeze-fracture analysis, presumably contain the water permeability channels induced by vasopressin [132–135]. All colchicine experiments, however, must be interpreted with caution, given the multiple effects of this drug on various cell functions [1, 131].

2. Surface polarity of epithelial cells. Epithelial cells exhibit a striking polarization of their plasma membranes into two domains—apical or luminal, and basolateral or contraluminal. The segregation of different enzymes and transporting systems between these two surface domains (Table II) accounts for the vectorial functions of epithelia, such as the undirectional transport of fluid, electrolytes, and nutrients and the secretion of specific cellular products into one of the two extracellular spaces that the epithelia separate.

The apical membrane, facing the "external" medium (eg, lumina of respiratory, reproductive, and gastrointestinal tracts, kidney tubules, and glandular acini), is typically free, and may exhibit morphological differentiations such as microvilli (in absorptive cells) which may become organized into a "brush border," and cilia [114]. Secretory vesicles in exocrine cells fuse only with this aspect of the plasmalemma [1, 7, 8, 34]. On the other hand, the basolateral surface faces the internal medium and specializes in adhesion to the basal lamina, possibly through specific receptors [150] and to other cells and, hence, is the site of the intercellular junctions (tight junctions, desmosomes, gap junctions). Its main secretory role may be the release through this face of basal lamina materials, such as collagen type IV [151]. The basolateral surface is comparable to the plasma membrane of nonepithelial cells, since it relates to the internal medium and, through it, with the blood stream, and possesses the transport systems necessary for cell survival. In

TABLE II. Markers of Apical and Basolateral Surfaces

Markers	Epithelium	References
Apical surface		
Alkaline phosphatase	I K L	136, 137
Leucine aminopeptidase	I K	136, 138
Oligo- and disaccharidases (sucrose-isomaltase, trehalase, lactase, maltase, glucoamylase)	I	139
cAMP-dependent protein kinase	K	140
Na^+-dependent transport of hexoses, amino acids, dipeptides, protons, sulphate, phosphate, chloride, lactate, ascorbate, bile acids	I K	141
Na^+-independent transport of fructose, P-aminohippurate, Cl^-	I K	141
ATP-dependent transport of H^+ and CO_3H^-	I K	141
Secretory component, IgG receptors	I	103, 104, 148
Basolateral surface		
Na^+K^+ ATPase	STE	142, 145
Ca^{++} ATPase	I K	141
Histocompatibility antigens (H_2)	I	146
Asyaloglycoprotein receptor	L	147
Secretory component, IgG receptors	I EG	103, 104, 148
Acetylcholine receptors	EG	149
Polypeptide hormone receptors	EG I K	140, 149
Adenylate cyclase	K	136, 140
Collagen, laminin binding sites	C	150
Na^+-dependent transport of amino acids, Ca^{++}, Cl^-	I K	141
Na^+-independent transport of neutral amino acids, glucose, lactate	I K	141

Abbreviations: I = intestine; L = liver; K = kidney; STE = sodium-transporting epithelia; EG = exocrine glands; C = corneal epithelium.

addition, it displays the hormone receptors for activation of the specific epithelial functions. These, which define a given epithelium, are usually carried out by special transporting and enzymatic systems in the apical surface.

In addition to their different protein composition, it is possible that apical and basolateral surface domains may differ considerably in their carbohydrate and lipid complement. Thus, Maylie-Pfenninger and Jamieson [152] described higher densities of wheat germ agglutinin, lotus lectin, and RCA I binding sites in the apical surface of isolated pancreatic cells from newborn and weaning rats. However, cells isolated from adult animals exhibited a homogeneous distribution of lectin binding sites which, the authors speculated, might be an artifact resulting from the redistribution of surface components upon dissociation (see section III,E). A lectin specific for sialic acid, limulin, binds preferentially to the apical surface of pancreatic acinar cells [153]. In contrast, in intestinal and kidney tubular cells, sialic acid is absent from various apical enzymes [139] but is present in basolateral proteins, like Na^+K^+ATPase (see section III,E).

There is also evidence indicating that apical and basolateral membranes have different lipid composition [154, 154a, 155, 155a]. Purified brush-border fractions from rat or mouse intestinal epithelial cells exhibit a lower content of lipids and are richer in cholesterol and glycolipids but poorer in phospholipids than basolateral membrane fractions. The molar ratio of cholesterol, phospholipid, and glycolipid was found to be 1:1:1 in microvillus membranes, as compared to 1:2.5:0.3 in basolateral membranes. Phosphatidyl ethanolamine (PE) and phosphatidyl choline (PC) were the major phospholipids in both membrane fractions, with PE predominating in the brush borders and PC in the basolateral membranes. Differences between the two surfaces in the molar ratios of fatty acids, as well as in the glycolipid composition, were also detected. The differences in lipid complement appear to be responsible for the higher fluidity of the basolateral membrane, as determined by steady-state fluorescence polarization [156] and fluorescence photobleaching recovery [156a], using fluorescent lipid probes. These experiments indicate that epithelial cells not only possess mechanisms to segregate proteins, but also lipids. In fact, Dragsten et al [156a] have recently shown that fluorescent lipid probes added to one of the two sides of epithelial cells in culture may redistribute to the other side, depending on their ability to flip-flop from the outer to the inner leaflet of the bilayer.

The boundary between the two regions of the plasmalemma is given by the tight junctions or zonulae occludentes, a continuous belt which appears in freeze fracture as a complex network of strands on the P face and grooves on the E face just at the limit between the apical and the lateral surface [157–159]. A discussion of the putative role of tight junctions in keeping the apical and basolateral molecular components separated is presented in section III,F.

B. Biogenesis of Some Epithelial Plasma Membrane Proteins

Several plasma membrane proteins of epithelial cells have been studied in relative detail in terms of their mode of association with the plasma membrane. These include two proteins of the intestinal brush border: sucrase-isomaltase complex [139, 163, 164] and aminopeptidase [138, 162], and the basolateral Na$^+$K$^+$ATPase typical of sodium-transporting epithelia [142–145]. The first two proteins are absent in the basolateral plasma membrane and can be solubilized from the brush border by treatment with proteases or with neutral detergents [138, 139, 160, 161]. The detergent-solubilized forms of the enzymes are slightly larger than the protease forms, the difference being given by a small hydrophobic peptide that remains attached to the membrane after proteolytic digestion [138, 139, 162–164]. This peptide provides the anchorage point to the membrane, and, differently from the plasma membrane proteins structurally related to glycophorin (H2 and HLA antigens, VSV and semliki forest virus glycoproteins), it is located at the amino terminal end of the molecule [162–164]. Aminopeptidase is probably a dimer (subunits of 130,000) in pigs, but perhaps only a monomer in rabbits [138]. Sucrase-isomaltase is a dimer with similar subunits of 140,000 daltons; one splits isomaltose and maltose and contains the hydrophobic attachment peptide, the other hydrolyzes sucrose and maltose and occupies a more peripheral position with regard to the plane of the bilayer [163, 164].

Studies on the biogenesis of sucrase isomaltase have shown that this protein is synthesized as a large molecular weight precursor, approximately twice the size of the individual subunits [165]. On the basis of pulse-chase analysis of subcellular fractions, Hauri et al [165] have postulated that this precursor is found initially (15′ after pulse) associated with Golgi membranes, and later (30′ after pulse) with basolateral membranes, before its arrival into the apical plasma membrane, where it is cleaved into the mature protein subunits by a pancreatic elastase in the lumen of the intestine. This cleavage could account for the locking of the enzyme complex in the apical surface [165]. Microtubules, which are concentrated in the Golgi areas and apical cytoplasm of intestinal cells [166], may possibly play a role in this transfer, since colchicine largely inhibits the migration of ^{3}H-fucose-labeled glycoproteins to the apical surface, but not their incorporation into basolateral membrane fractions [167]. All these experiments should be interpreted with extreme caution, given the problems involved in obtaining pure plasma membrane fractions, especially those of basolateral origin [154, 154a, 155, 155a]. Indeed, a contamination of the latter with Golgi vesicles could explain these results. A similar biogenetic pathway has been postulated for secretory component (see section II,D), a membrane-bound glycoprotein which mediates the endocytic, basal to apical, transfer of dimeric IgA and polymeric IgM across several epithelia [103, 104]. This protein, however, differently from the brush-border enzymes mentioned above, is detected in both the apical and basolateral surfaces of intestinal cells [104]. Like sucrase-isomaltase, secretory component is synthesized as a large molecular weight precursor, and is also cleaved by proteases upon arrival to the apical plasma membrane [105]. More work is clearly needed to establish the validity and generality of these mechanisms for the biogenesis of apical proteins in epithelia.

Na^+K^+ATPase, a typical basolateral enzyme in transporting epithelia [142–145] is composed of two nonidentical subunits: the large, alpha chain (MW 120,000), nonglycosylated and relatively hydrophobic; and the small, beta chain (MW 55,000), a sialoglycoprotein [168, 169]. The large chain is structurally related to two other transporting enzymes, Ca^{++}ATPase and H^+K^+ATPase [170], is the phosphorylated intermediate associated with catalysis, and has several membrane-associated regions [170]. Both subunits are thought to span the lipid bilayer [171]. In the membrane, protein units interact to form dimers (α_2, β_2), as has been shown for several other membrane proteins [172]. Immunocytological localization of Na^+K^+ATPase on ultrathin frozen sections has revealed the enzyme on the apical surface of kidney proximal and distal tubules, although in much smaller amounts than on the basolateral regions [143, 144]. Recent work indicates that the small subunit is synthesized by membrane-bound polysomes as a precursor with a smaller molecular weight, which is cotranslationally glycosylated and presumably contains an uncleaved signal sequence; the large subunit, instead, is apparently synthesized by free cytoplasmic polysomes [173].

The biogenesis of mouse H2 antigens, which are basolateral proteins in intestinal epithelial cells [146], has also been studied, although in nonpolarized cell lines of lymphatic origin [72]. The human counterpart, HLA histocompatibility antigens, have a very similar structure and biogenetic mechanisms, which generally parallel those described for the VSV G protein [73] (see section II,C). It remains to be established whether the structural features and biogenetic mechanisms described for H2 and HLA antigens in nonpolarized cells remain the same in polarized epithelial cells.

The different structures, functions, and biochemical compositions of the two epithelial plasmalemmal domains justify their consideration as two separate cell organelles. It may be assumed that the sets of apical or basolateral proteins share common structural information (or sorting out signals) and common biogenetic mechanisms that determine their final surface localization. Understanding the nature of these mechanisms and signals should provide valuable information to predict how the sorting out of components destined to different cellular membranes operates.

An excellent tool for the study of this important biological problem has recently become available with the development of cell lines of epithelial origin, which preserve under tissue culture conditions polarity properties from natural epithelia.

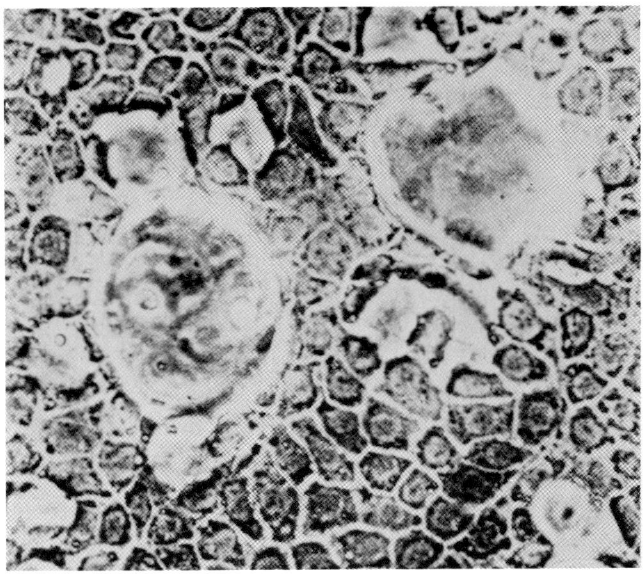

Fig. 2. Domes in a confluent monolayer of MDCK cells. × 480.

C. Polarized Epithelial Monolayers In Vitro

When grown to confluency in tissue culture, some cell lines of epithelial origin develop hemicystic structures or domes, owing to the accumulation of fluid and electrolytes between the monolayer and the Petri dish (Fig. 2). This results both from unidirectional, ouabain-sensitive transport of salts and water and from the existence of functional tight junctions—two typical properties of natural transporting epithelia [174]. Dome formation, which with time-lapse cinematography resembles the "gentle boiling of oatmeal," was originally reported for MDCK cells [175], a cell line derived from dog kidney by Madin and Darby in 1958 [176]. It has since become evident that many cell lines of diverse epithelial origin form epithelial-like monolayers with morphologically well differentiated apical and basolateral surfaces, and intercellular junctions [see 177 for a review].

When grown on a permeable substratum, such as Millipore filters [178] or collagen-coated nylon nets [179], MDCK cells develop transepithelial electrical resistances of approximately 100 ohms cm^2, a small transepithelial electrical potential (about 1 mV, positive on the basal side) and cationic selective permeability—Na$^+$ is preferred ten times over chloride [179]. Electron microscopy reveals typical epithelial differentiations, such as microvilli on the apical surface and tight junctions and desmosomes on the lateral surface (Figs. 3, 4). The presence of tight junctions in freeze fracture correlates well with the ability to develop transepithelial electrical resistance and domes; fibroblastic cell lines show no evidence of either of them [179]. Development of polarity does not depend on the interaction with a specific substratum, since it occurs on plastic, glass, collagen, or nitrocellulose filters [178, 179]. MDCK cells exhibit properties of distal convoluted kidney tubules as evidenced by functional and morphological studies [180] and by experiments with monoclonal antibodies raised against their surface [181]. They grow in monolayer culture, proliferate with a doubling time of approximately 1 day, and can be cultured in hormone-supplemented, serum-free medium [182]. Monolayers of MDCK cells are being used as a model system to study the transepithelial migration of leukocytes [183].

The polarized distribution of plasma membrane proteins in MDCK cells has been demonstrated by lactoperoxidase-catalyzed iodination of monolayers grown on Millipore filters [184]. Furthermore, the basolateral surface exhibits higher binding of ^{3}H-ouabain and increased density of intramembranous particles [185, 186]. Using immunofluorescence, Louvard [187] has localized leucine aminopeptidase to the apical plasma membrane and Na$^+$K$^+$ATPase to basolateral membranes.

The following section describes a model system recently developed to study the mechanisms of polarization of plasma membrane proteins in epithelial cells. The system involves the infection of polarized epithelial cell lines with enveloped RNA viruses. Because of the successful use of the glycoproteins of these viruses in studies of plasma membrane biogenesis, it was of interest to determine whether they would also be useful tools for similar studies in epithelial cells—ie, whether they behave as intrinsic epithelial plasma membrane proteins by becoming segregated into one of the two plasma membrane domains upon infection of polarized epithelial cell lines.

D. Polarized Assembly of Enveloped Viruses in Epithelial Cells

When MDCK cells are infected with enveloped RNA viruses, a striking asymmetry of viral budding is observed [188]. Influenza (WSN strain, a myxovirus) sendai, and simian virus 5 (two paramyxoviruses) are selectively

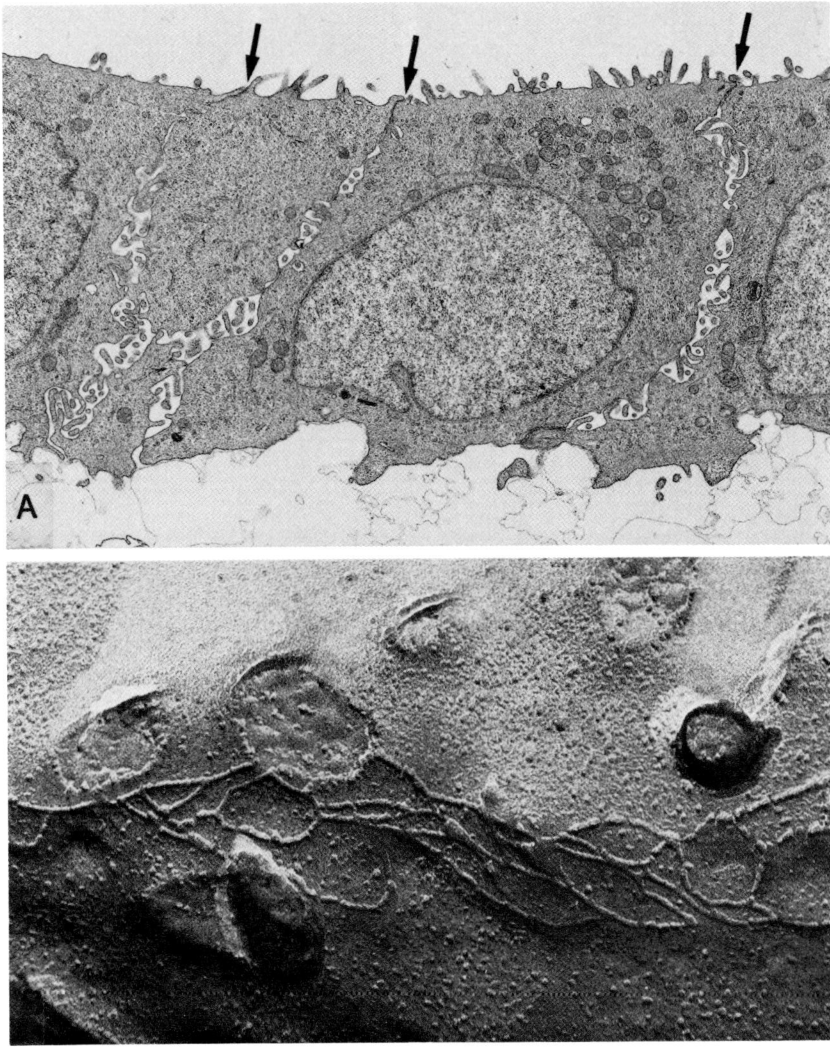

Fig. 3. Confluent monolayer of MDCK cells. A) Thin-section electron microscopy of a monolayer grown on a Millipore filter. Note the microvilli on the apical surface and tight junctions (arrows) separating apical and basolateral aspects of the plasma membrane. × 5,500. B) Freeze-fracture replica of a confluent MDCK monolayer. Note the typical tight junctional strands on the P face. × 63,000. Reprinted with permission from Cramer et al [183].

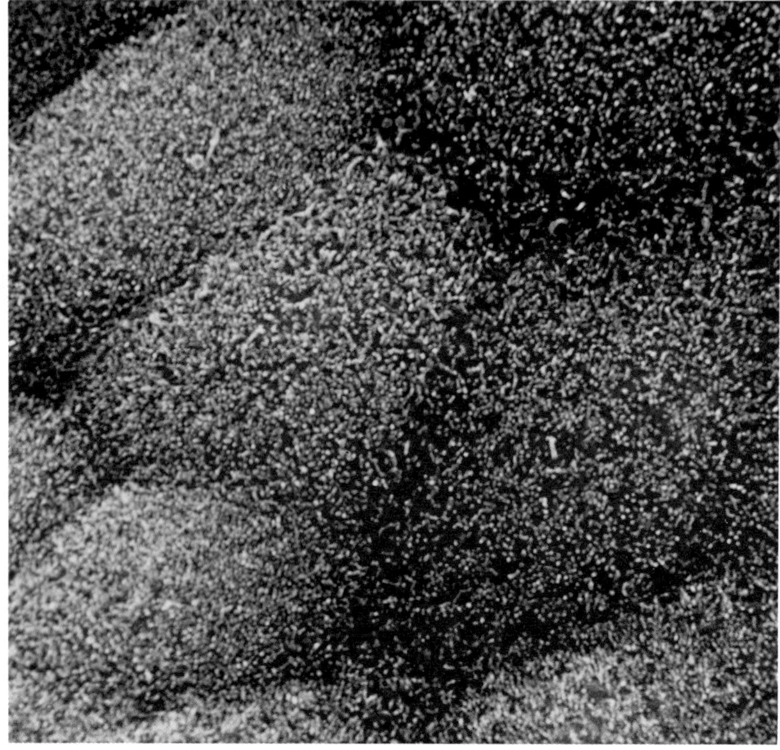

Fig. 4. Scanning electron microscopy of MDCK monolayers ($\times$ 1,400). Reprinted with permission from Ojakian [231].

assembled from the free (or apical) cell surfaces (Figs. 5, 6). On the other hand, VSV, a rhabdovirus, buds exclusively from the basolateral plasma membrane (Fig. 5). Infection with murine leukemia viruses (MLV) results in persistent infection of MDCK cells with selective budding from the apical cell surface (Rodriguez-Boulan et al, unpublished results). Thus the direction of budding is a specific property of each virus, since different viruses express opposite polarities in the same cell line and each virus retains the same polarity in several other epithelial cell lines (such as MDBK, derived from bovine kidney [188] and a rabbit tracheal epithelial line (unpublished results)). Infection of nonpolarized cells, such as chick embryo fibroblasts, results in budding from both the free and the attached cell surface, indicating that asymmetric viral budding is a property unique to polarized (epithelial) cells.

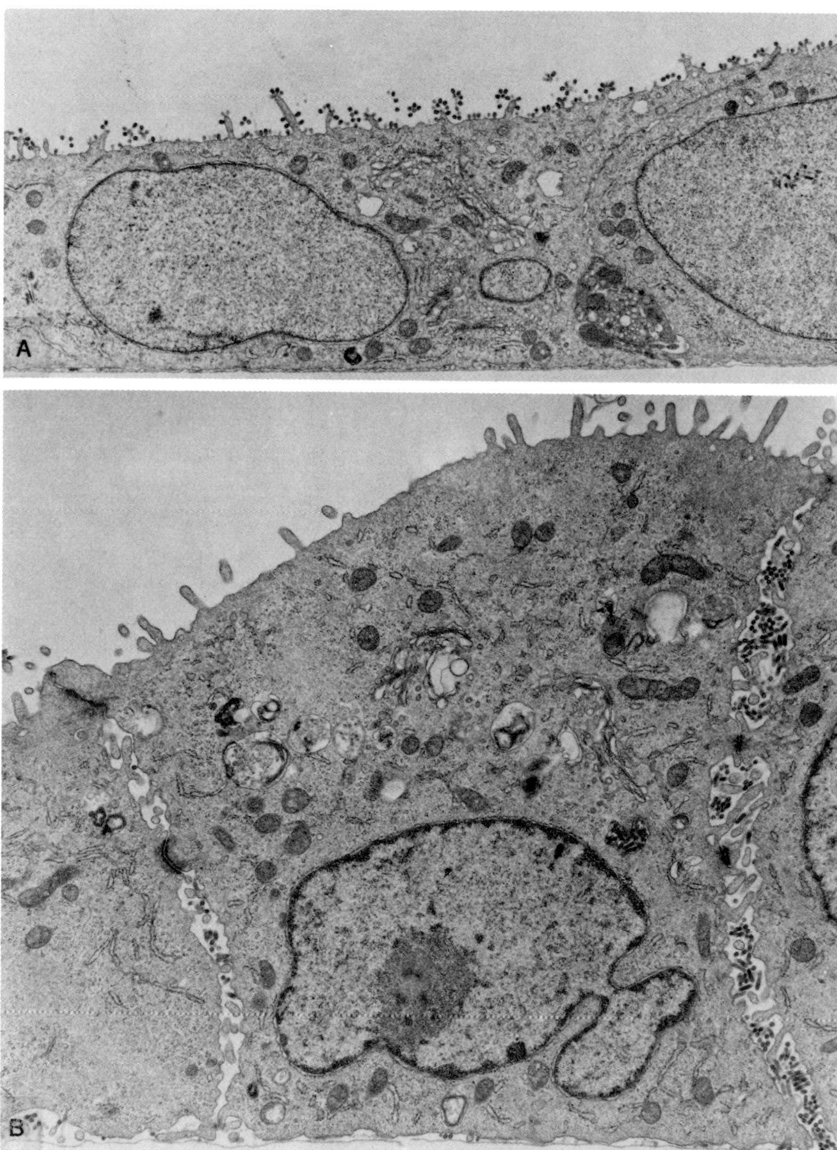

Fig. 5. Asymmetric budding of influenza and vesicular stomatitis virus from MDCK cells. Confluent monolayers of MDCK cells were infected with influenza (A) or VSV (B) at a multiplicity of infection of 10 pfu/cell. The monolayers were fixed with 2% glutaraldehyde after 8 hours of infection and processed for electron microscopy. A) Note the budding of influenza virus from apical regions of the plasma membrane. Also, note the presence of several Golgi complexes (× 12,000). B) Budding of VSV from the basolateral surface of MDCK cells (× 16,500). Two supranuclear Golgi apparatus are clearly visible.

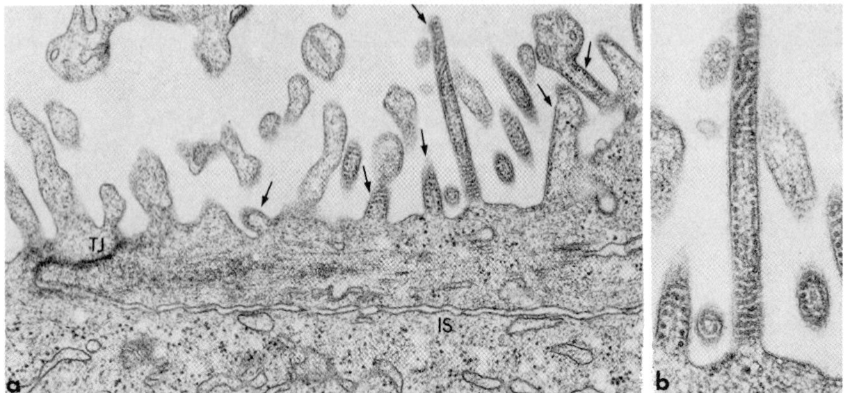

Fig. 6. Apical budding of SV5 from MDBK cells. Note the absence of virions at the lateral cell surface (× 38,000). Reprinted with permission from Rodriguez-Boulan and Sabatini [188].

There are some reports of polarized budding in natural epithelia. Murphy and Bang [189] and Bang [190] observed budding of influenza and Newcastle disease virus from the free surface of the chorioallantoic membrane of chick embryos. It has been reported that nuclear polyhedrosis baculoviruses bud exclusively from the basal aspect of insect intestinal cells [191]. Pickett et al [192] have reported budding of mammary tumor virus from the apical surface of dissociated mouse mammary adenocarcinoma cells grown in tissue culture. Polarized viral budding, therefore, appears to be a widely distributed property of epithelial cells, and may play an important role in the spread of viral infection in the organism.

It has been shown that polarized viral budding is preceded by the segregation of viral envelope proteins in the surface domain the virus utilizes for budding [193], (Figs. 7, 8). However, the viral nucleocapsids do not appear to show any preferential cytoplasmic localization close to the budding surface [193]. The implications of these findings are twofold: 1) The polarized distribution of viral glycoproteins appears to be the main determinant of polarized viral budding. The role of the M (matrix) protein in this process, however, has not yet been determined. 2) Viral envelope glycoproteins share with intrinsic epithelial surface proteins the ability to become segregated into specific surface domains. This implies that, during their biogenesis, some structural features common to the glycoproteins of sendai, influenza, sv5, and retroviruses are recognized by cellular distribution mechanisms of epithelial cells as a signal for delivery to the apical surface. Similarly, the VSV G protein must be seen as a basolateral protein and is therefore directed to that domain.

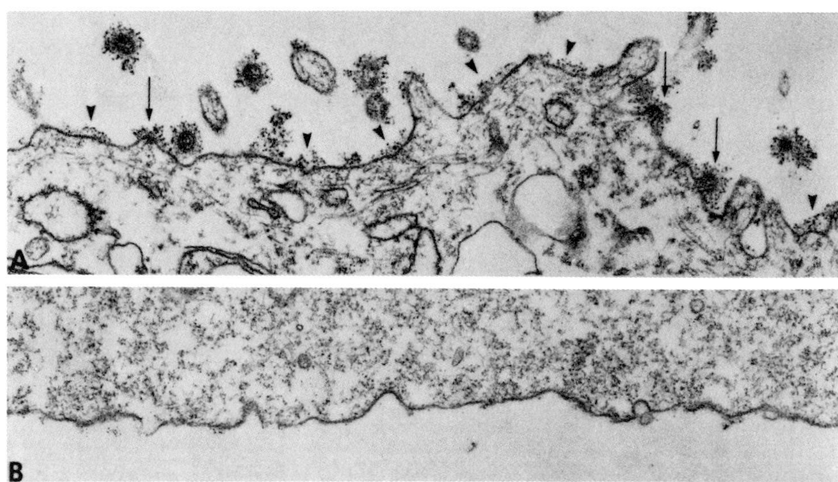

Fig. 7. Localization of influenza virus envelope proteins in MDCK cells by immunoelectronmicroscopy. The monolayers were fixed with 4% paraformaldehyde after 6 hours of infection, removed from the Petri dish by scraping, and incubated with rabbit antibodies against influenza virus, followed by treatment with ferritin-goat antirabbit conjugates ($\times$ 43,000). A) Apical region of infected cells. Note patches of envelope proteins (arrowheads) and budding virions (arrows). B) Basal plasma membrane of infected cell. Reprinted with permission from Rodriguez-Boulan and Pendergast [193].

The experiments indicate that, indeed, the glycoproteins of enveloped RNA viruses are good model systems to study the mechanisms that determine the polarized distribution of intrinsic plasma membrane proteins in epithelial cells. Since a feature common to all of them is their glycoprotein nature, experiments were carried out to determine the role of glycosylation in determining asymmetric surface distribution.

E. Glycosylation and Epithelial Polarity

Viral glycoproteins and many membrane and secretory glycoproteins contain one or more carbohydrate side chains linked via N-glycosydic bonds to asparagine residues in the polypeptide [4, 13, 14, 32, 33]. The oligosaccharides are typically of two kinds: simple and complex [13, 14]. The first type is composed of 8–9 mannose and two N-acetylglucosamine residues. Complex chains exhibit a "core," consisting of 3–5 mannose and two N-acetylglucosamine residues and two or three "branches," given by the trisaccharide N-acetylglucosamine-galactose-sialic acid. Fucose may be found attached to one of the core N-acetylglucosamines. The biosynthesis of these carbohydrate moieties, elucidated in the last few years, has been found to

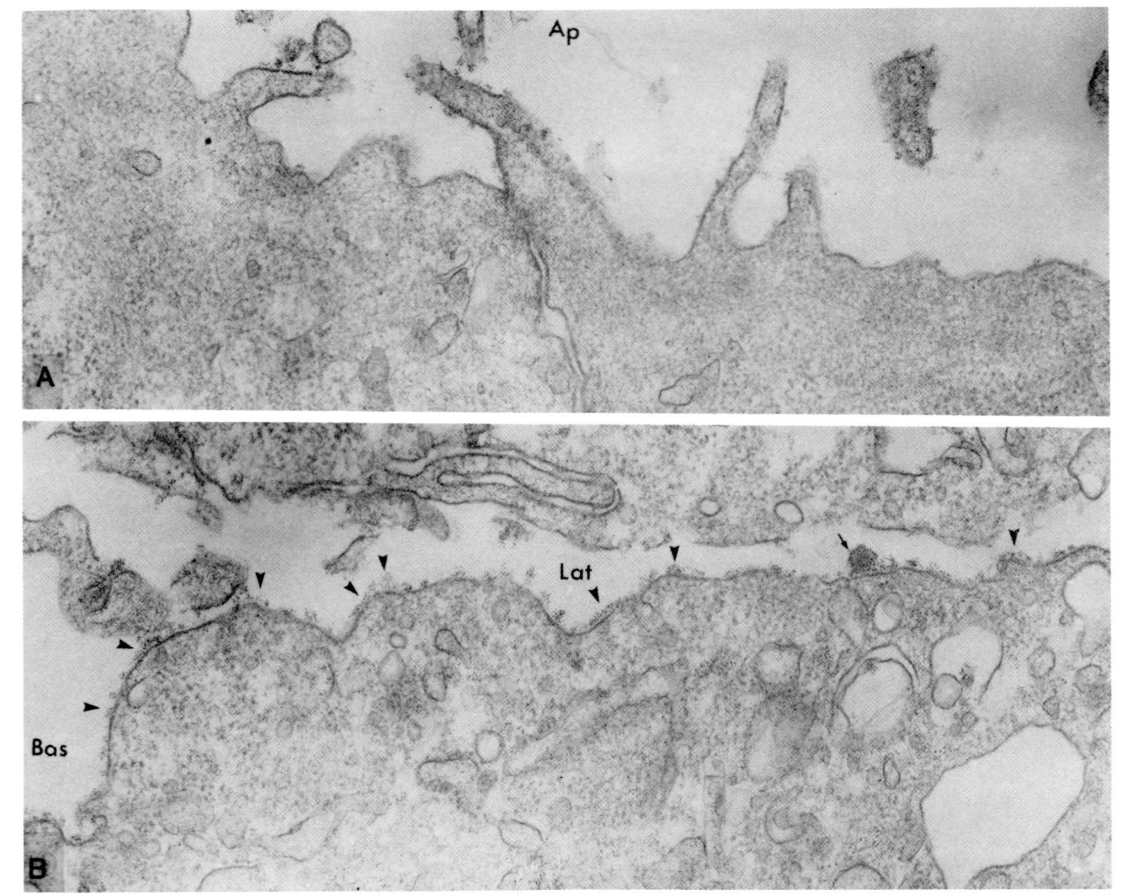

be a complex multienzyme process that is initiated by the transfer in block from a lipid intermediate to the nascent glycoprotein of an oligosaccharide consisting of two N-acetylglucosamine, nine mannose, and three glucose residues [13, 14]. This oligosaccharide is "trimmed" by removal of the glucose and several mannose residues to the size of the core found in complex side chains or, alternatively, loses only the glucose residues and remains as a simple side chain [62–64]. The branch carbohydrates are added sequentially by specific transferases [13, 14]. The core transferase and some of the trimming enzymes (those for glucose) are located at the endoplasmic reticulum, whereas the mannose trimming enzymes and the branch sugar transferases are localized in the Golgi apparatus [13, 14].

The functional role of glycosylation in the intracellular transport of membrane and secretory proteins is controversial. The migration of some proteins appears to be highly sensitive to the addition of glycosylation inhibitors, but others do not appear to be affected [87, 194, 195].

It was noticed [188] that the glycoproteins of several apically budding viruses as well as several apical glycoproteins from natural epithelia typically lack sialic acid in their composition, whereas, in contrast, basolateral proteins such as the G protein of VSV and $Na^+K^+ATPase$ display sialic acid in their carbohydrate chains [168]. Because of this suggestive correlation, we investigated the role of carbohydrates in determining the polarized distribution of viral glycoproteins, using three different approaches [196]: 1) development of lectin-resistant mutants of MDCK cells with defects in the synthesis of their glycoproteins [197]; 2) use of tunicamycin, an antibiotic that abolishes the glycosylation of viral glycoproteins [198]; and 3) use of a temperature-sensitive mutant of influenza WSN which, at the nonpermissive temperature (39°C), retains sialic acid because of a defective viral neuraminidase [199].

VSV or influenza glycoproteins, synthesized in mutant MDCK cells resistant to levels of concanavalin A (Con A^R) or Ricin (RCA^R) toxic for the wild-type cells, exhibited characteristically increased electrophoretic mobility. The genetic defect in RCA^R cells probably resulted in the absence of terminal di- or trisaccharides, since there was no observable incorporation of ^3H-galactose into G protein upon infection with VSV [197]. Many cellular

---

Fig. 8. Localization of VSV glycoprotein in MDCK cells by immunoelectronmicroscopy. The monolayers were fixed with 4% paraformaldehyde after 6 hours of infection, scraped from the Petri dish, and treated with biotinized sheep antibodies against VSV followed by incubation with ferritin-avidin conjugates [189]. Note the heavy labeling of the basolateral surface (B arrowheads) and around a virion (arrow). The apical surface (A) exhibits a much lower amount of ferritin binding. Some ferritin molecules are found in apical regions close to the tight junction, presumably because of junctional disruption as a consequence of infection [189]. ×45,000.

glycoproteins, which were biosynthetically labeled with ^{3}H-glucosamine, also had decreased apparent molecular weights in gels, suggesting constitutive changes in glycosylation. In spite of their altered carbohydrate moieties, the mutant cell lines retained structural and functional polarity as determined by their ability to develop tight junctions and to form domes on impermeable substrata [197]. Furthermore, viral budding polarity was preserved and, as in wild-type cells, the viral glycoproteins were distributed asymmetrically on the cell surface, maintaining the correct orientation: influenza glycoproteins accumulating on apical and VSV G protein on basolateral regions of the plasma membrane [196] (Fig. 9).

Likewise, the complete inhibition of glycosylation by tunicamycin treatment, or the retention of sialic acid moieties by the hemagglutinin in the influenza ts mutant, did not affect the polarized distribution of the viral glycoproteins or the asymmetric viral budding. The amount of protein reaching the surface and the yield of infective viruses, however, were reduced in the presence of tunicamycin [196]. This effect was considerably more pronounced for influenza viruses than for VSV. Roth et al [200] have also reported that tunicamycin does not affect polarized viral budding in MDCK cells. It can therefore be concluded that carbohydrates are not components of the sorting signals that direct viral glycoproteins (and possibly cellular proteins) to specific plasma membrane regions in epithelial cells.

The results of these experiments are in agreement with an increasing body of evidence indicating that the effect of carbohydrate removal on the biological properties of glycoproteins is highly variable and depends on the proteins themselves. The intracellular migration of some secretory proteins, for example, is largely inhibited by tunicamycin, whereas other glycoproteins remain totally unaffected under the same treatment [see 87, 194, 195 for examples].

F. Endogenous and Exogenous Determinants of Polarity: The Role of Tight Junctions and Cell-Substrate or Cell-Cell Interactions

The relevant questions related to this point are 1) What factors determine the initial polarization of the epithelial plasma membrane into apical and basolateral regions? In other words, how is polarity established? and 2) Once polarity is established, how is it maintained? What is the role of the tight junctions? Why are tight junctions always located at the apical end of the intercellular space in epithelia? Is polarity reversible?

Tight junctions (zonulae occludentes) form a continuous "belt" at the boundary between apical and lateral surfaces of secretory and absorptive epithelial cells [157–159]. At the level of the tight junctions, the outer leaflets of the membranes of two or three contiguous cells appear to fuse, as seen by thin-section electron microscopy, which suggests physical continuity between the membranes of these cells. This point, however, has not been

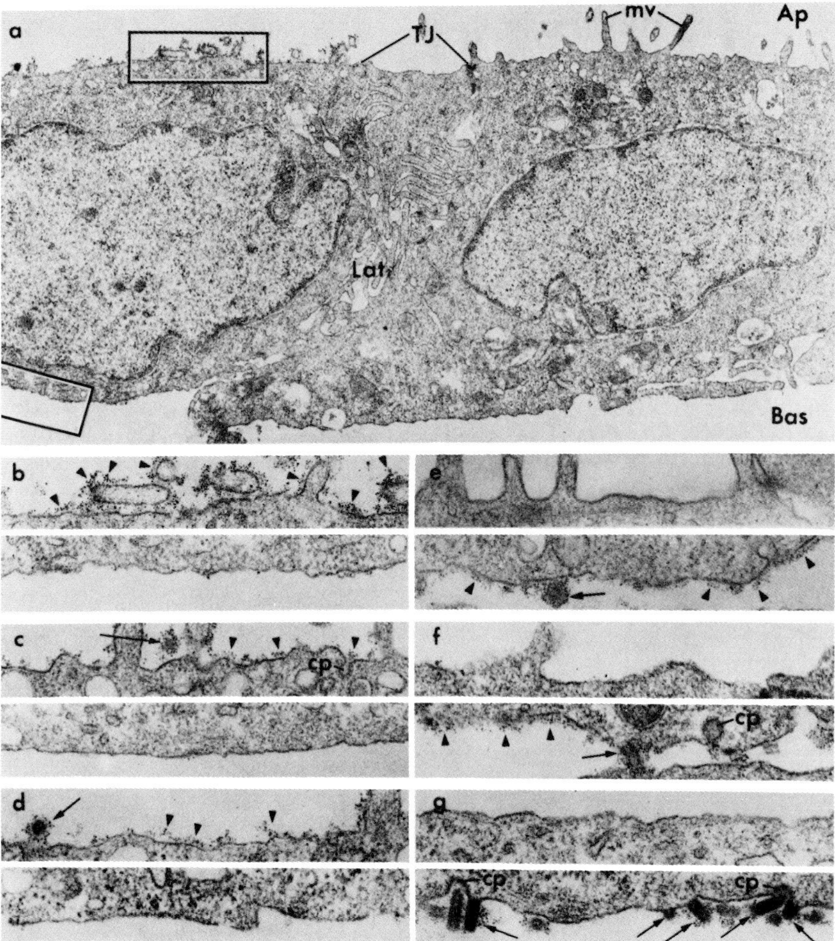

Fig. 9. Polarized distribution of viral glycoproteins in lectin-resistant MDCK cells. Confluent monolayers of wild-type, RCA^R, and Con A^R MDCK cells grown on 35-mm plastic dishes were infected with VSV or 3.5–5 hours after infection by VSV, or 6–8 hours after infection by WSN, monolayers were fixed in paraformaldehyde (4% in PBS, 20 min, rt), scraped of the substratum, and incubated with biotin-conjugated antiviral IgG (50 μg/ml in PBS, 30 min, rt), followed, after several washes in PBS, by avidin-conjugated ferritin (1 mg/ml in PBS, 1.5 h, rt). Samples were washed in PBS, three times; 0.1 M Na cacodylate, pH 7.4, once; fixed in glutaraldehyde (2% in cacodylate); and processed for thin-section electron microscopy. The boxed areas in (a) are shown at higher magnification in (b). Each pair of figures represents the apical (top) and basolateral (bottom) region of a single cell. Arrows indicate budding or released virions; arrowheads point at patches of ferritin on the plasma membrane. Ap, apical plasma membrane; Bas, basal plasma membrane; Lat, lateral plasma membrane; TJ, tight junctional areas; mv, microvilli; cp, coated pits. a) wt and WSN (× 14,500); b) wt and WSN; c) RCA^R and WSN; d) Con A^R and WSN; e) wt and VSV; f) RCA^R and VSV; g) Con A^R and VSV (b–g, × 43,700). Reprinted with permission from Green et al [196].

demonstrated functionally. It is well established that tight junctions constitute an intercellular diffusion barrier for ions and macromolecules [201]. This is reflected in the generation of a transepithelial electrical resistance, the level of which determines the arbitrary classification of epithelia into "leaky" and "tight," with the cutoff point at around 100–500 Ω cm^2 [159, 201]. Furthermore, the strategic location of tight junctions suggests that they may play an important role in keeping the components of the apical and basolateral membranes separated. Pisam and Ripoche [202] have reported the redistribution of apical markers over the entire cell surface in dissociated toad bladder epithelial cells. In agreement with these findings, the brush-border enzymes leucine aminopeptidase and alkaline phosphatase are localized over the entire cell surface of isolated intestinal cells [203]. Meldolesi et al [204] have reported that incubation of guinea pig pancreatic lobules in Ca^{++}-free Ringer containing 0.5 mM EGTA results in the progressive fragmentation of the occluding zonulae with formation of multiple discrete junctions (fasciae occludentes) localized in the lateral and luminal plasmalemma. This results in the disappearance of the heterogeneity in density of intramembrane particles on the P-fracture faces of the basolateral and lumenal plasmalemma (the normal densities are about 2800 and 800 particles/μ^2 respectively [204]. MDCK cells also exhibit higher density of intramembrane particles on the basolateral surface [185, 186]; as in pancreatic cells, this difference is abolished by incubation in the presence of Ca^{++}-free medium containing EGTA [185].

In order to understand the relationship between tight junctions, substrate attachment, and epithelial polarity, experiments were carried out involving the infection of dissociated MDCK cells with enveloped viruses [207]. After treatment with trypsin-EDTA, the isolated MDCK cells were kept in suspension culture overnight to allow for resynthesis of cell surface proteins. The following day the aggregated cells were removed by filtration and the single cells were infected with either VSV or influenza, which bud from opposite surfaces in confluent monolayers. Some cells were kept in suspension after infection; others were plated sparsely on collagen gels to prevent cell-cell interaction. The following results were obtained:

- Isolated cells in suspension exhibited viral budding from the entire cell surface indicating overall loss of polarity.
- Isolated, attached cells exhibited the correct budding polarity for VSV or influenza (influenza budded from the free surface while VSV was produced from attached regions).
- Clumped, suspended cells, with no substrate interaction, again showed the correct polarity of budding for both viruses.

These experiments clearly demonstrate a role for the interaction of the epithelial cell surface with a substrate or with other cells in the establishment of polarity. Although more experiments are needed to determine to what extent the observed polarity of viral budding from isolated, attached cells represents the situation with intrinsic epithelial surface proteins, it appears that the formation of intercellular junctions is not essential for this process. Tight junctions may thus represent a consequence of the establishment of polarity and cell-cell interaction rather than its cause.

It may be speculated that the interaction with a substrate results in the recruitment of a group of "sticky" attachment sites such as those described in some fibroblastic and epithelial cell lines [205, 206] into the adherent membrane surface, which triggers the segregation of apical and basolateral regions. Such a phenomenon has been described for Fc receptors in macrophages, which are segregated on the adherent region of cells plated on a surface saturated with IgG [208]. In the organism, epithelial cells interact with the basal lamina and with other epithelial cells. Sugrue and Hay [150] have recently reported that corneal epithelium removed from underlying extracellular matrix experience an active process of blebbing from the normally smooth basal surface. Blebbing is stopped by incubation of the corneal epithelia in the presence of soluble collagen (of type I, II, or III), laminin, or fibronectin, but not by albumin, IgG, glycosaminoglycans, or attachment to glass. These experiments suggest the existence of relatively nonspecific collagen receptors on the basolateral surface. Goldberg [209] has described collagen receptors on the cell surface of fibroblasts, but failed to find them in epithelial cells. This may not represent a lack of collagen receptors, however, since the basolateral surface of epithelial cells is not readily accessible to macromolecules because of the presence of tight junctions. Rubin et al [210, 211] have recently hypothesized that the attachment of rat hepatocytes to collagen, a fibronectin-independent process [212], is mediated by receptors that recognize multiple repeating sites (eg, Gly-Pro-Hyp) along the collagen molecule. Since the putative collagen receptor has not yet been isolated, it is not possible to distinguish between two different possibilities that may contribute to the polarization of epithelial surface proteins: 1) Proteins in the basolateral surface of epithelial cells possess the ability to recognize the attachment sites in the basolateral membrane, and thus are segregated into that surface; 2) all proteins in the basolateral surface have "sticky" properties—ie, the ability to interact with a substrate and therefore to be polarized into the adherent surface.

1. Development of tight junctions in cultured epithelial cells. The availability of epithelial cell lines has made it possible to study the development of tight junctions in detail. Cereijido et al [179, 213] have studied this process by plating MDCK cells dissociated with trypsin-EDTA at high densities on collagen-coated nylon disks and following the development of transmono-layer electrical resistance (TER) as a function of time. The cells attached to the collagen, spread within 1 hour and, in 2–4 hours, a transepithelial resistance started to develop; it reached peak values of up to 340 Ω cm^2 in 20–24 hours, and by 48 hours had returned to a stable plateau of around 100 Ω cm^2. The development of TER was correlated with the appearance of tight junctions in freeze-fracture replicas [179, 213]. The development of selective permeability for sodium over chloride, typical of confluent monolayers, followed a slower time course, reaching a peak value 2 or 3 days after plating, which indicates that these two properties of tight junctions can be dissociated [213].

Confluent monolayers of MDCK cells, like the intestinal epithelium, exhibit an increased concentration of actin and α-actinin in regions close to the tight junctions, which gives a ringlike image in immunofluorescence localization experiments [187, 214]. Incubation in Ca^{++}-free medium or in the presence of tumor promoters have been reported to cause an increase in transepithelial conductivity owing to disassembly of the tight junctions [231, 232]. Meza et al [214] have found that incubation of confluent MDCK monolayers in the presence of cytochalasin B (but not colchicine) causes a drop in the TER, accompanied by a fragmentation of the actin ring, both processes being reversible. Ojakian [231] has also reported that cytochalasins B and D cause a decrease in the TER in confluent monolayers of MDCK cells, but, in contrast to Meza et al [214], he found a significant inhibitory effect of colchicine and vinblastin. In a recent work by Griepp et al [215], cytochalasin B was found to block the development of TER after heavy cell plating but did not decrease the conductivity of confluent monolayers of MDCK cells. In spite of these contradictory results, the combination of immunocytological, ultrastructural, and electrophysiological data strongly suggests that actin microfilaments are involved in the formation of tight junctions in epithelial cells.

The reason is not altogether clear why tight junctions are, in fact, located where they are: at the boundary between apical and lateral epithelial surfaces. This localization might reflect the existence of directional bulk movements of membrane material in the epithelial cell surface. DiPasquale, for example, has shown that carbon or latex particles, or Con A–treated red blood cells, move centripetally from the edge of marginal cells in corneal epithelial monolayers [216]. Centripetal movement of fluorescent Con A and viral envelope proteins has also been described [217, 218] in nonepithelial cell lines. Louvard [187] has shown that interiorization of leucine aminopeptidase in MDCK cells upon incubation with specific antibodies is followed by reinsertion of the enzyme in peripheral regions of the cell, and subsequent redistribution over the whole apical cell surface. We have found similar results with respect to the insertion of influenza's hemagglutinin in the apical surface of subconfluent MDCK cells (unpublished results).

In addition, the edges of free epithelial cells display a higher concentration of adhesive sites and active fluctuation, ruffling, blebbing, and microspike activity [216, 219]. If tight junctions would initially consist of point adhesions between the neighboring epithelial cells that later coalesce to form the strands, a contripetal movement of membrane on the free cell surface would explain their characteristic localization at the apical-lateral interface (Fig. 10). They

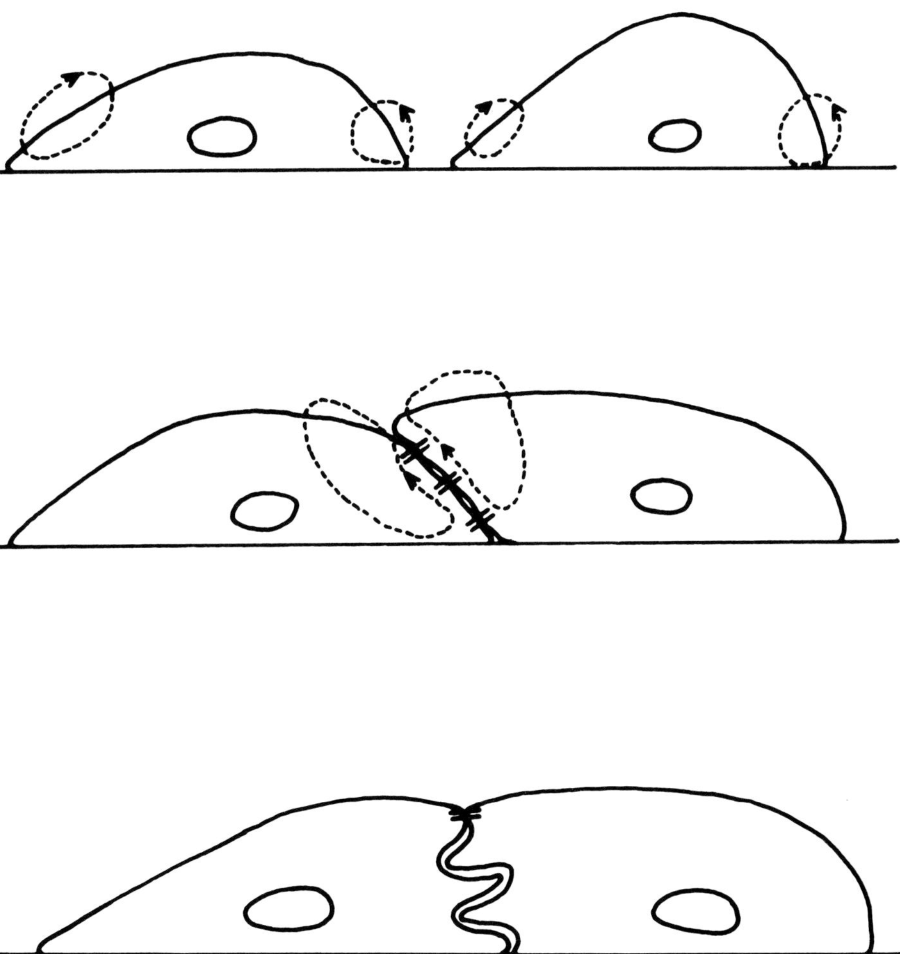

Fig. 10.   Hypothetical stages in the formation of tight junctions. Top) Epithelial cells, which are making no contact or incomplete contact with other cells, exhibit a centripetal movement of membrane material at their periphery (probably compensated by an intracellular centrifugal movement of membrane). Middle) Upon contact, punctual tight junctions form between the adherent surfaces. The centripetal movement of membranes pushes the incomplete tight junctional strands toward the apical end of the intercellular space, where they coalesce into a complete belt (bottom).

could be stabilized into this position by association with the characteristic cytoskeletal structures underlying the apical surface (section III,A).

2. Reversibility of epithelial polarity. Once the epithelial monolayer is formed, how stable is it? Can it be reversed? Experiments with hog or rat thyroid epithelial cells in culture have demonstrated that they can change their polarity according to the culture conditions [220–222]. When cultured in the presence of low serum concentrations(0.5%), the cells form monolayers on a Petri dish or cystic structures in suspension culture; these cysts, or "vesicles," have an orientation opposite to that found in the natural follicular epithelium; ie, the apical surface (with the tight junctions) is facing the outside medium. The Golgi complex is localized between the nucleus and the apical surface. Addition of thyroid stimulators (such as thyrotropin, prostaglandin E2, or dibutyryl cyclic AMP) or embedding in collagen gels causes the cells to reorient to form "follicles," with the same polarity as the natural epithelium (apical surface, tight junctions, and Golgi apparatus facing the lumen). It is not clear, however, whether other epithelial cell types show a similar ability to reverse their polarity. Not even thyroid cells in vivo display such a capacity [223].

However, cytoplasmic organelles may change their position without a reversal of the surface polarity features. For example, during two distinct periods in the development of the chick corneal epithelium, which correlate in time with the deposit of extracellular matrix beneath the epithelium, the Golgi complex changes its position from supra- to infranuclear in the basal cell layer [224, 225]. Similarly, the ameloblasts in the epithelial enamel organ excrete enamel matrix from their basal pole only after the Golgi apparatuses have shifted from apical to basal location [226]. The centrioles and their associated microtubules change their intracellular position along with the Golgi apparatus [225]. Given the important role of the Golgi complex in the late stages of secretion, the change in its cell position according to the polarity of secretion suggests a possible causal relationship between Golgi and polarity.

G. Intracellular Pathways of Apical and Basolateral Proteins: Mechanisms of Polarity

An important question in epithelial polarity is *where* apical and basolateral proteins are segregated in the cell. Are both classes of proteins inserted asymmetrically in the plasma membrane? Or are they randomly inserted and then segregated by specific mechanisms?

Using the model system of virally infected epithelial monolayers, we sought to determine the intracellular pathway followed by apically and basolaterally directed viral glycoproteins [227]. In order to study the pathways of both classes of glycoproteins simultaneously, we carried out experiments with MDCK monolayers double-infected with viruses of opposite budding polarity: SV5 and VSV, or influenza and VSV. Previous reports had indicated that it was possible to coinfect epithelial cell lines with SV5 and VSV and to obtain pseudotypes with mixed viral glycoprotein composition [228, 229] containing a predominance of VSV genomes.

Recent work by us [227] and by others [230] indicates that MDCK cells coinfected with apically (SV5 or influenza) and basolaterally (VSV) directed viruses preserve to a large extent viral budding polarity. Immunoelectron microscopy experiments with ultrathin frozen sections of monolayers of MDCK cells infected with influenza and superinfected with VSV, using colloidal gold particles of different size for simultaneous localization of the envelope glycoproteins of both viruses, indicate that both types of proteins share the intracellular pathway until, at least, the Golgi apparatus [227]. The available evidence does not allow to distinguish between two different possibilities: 1) Apical and basolateral glycoproteins are sorted into different post-Golgi vesicles that selectively fuse with the respective surfaces; and 2) apical and basolateral proteins are inserted into one of the two plasma membrane domains or at random over the entire cell surface, and are later sorted out by specific retrieval mechanisms.

Given the increasing amount of evidence on the participation of the Golgi apparatus in the sorting of various kinds of organellar proteins [77, 79], it is not unconceivable that different vesicles could be generated by this organelle, with ability to recognize apical or basolateral regions of the epithelial plasmalemma (Fig. 1). Candidates for this role could be different species of coated vesicles, since there is some evidence that these structures may be involved in the transport of G protein to the apical surface [66, 67]. However, the alternative possibility cannot be ruled out: Apical and basolateral proteins could be first inserted into one of the surfaces or randomly into both and then sorted by specific mechanisms. The steady-state level of protein in the surface opposite to the budding surface would still be low if the protein were quickly removed by the putative retrieval mechanisms. There is evidence that the intestinal enzyme sucrase-isomaltase is inserted in the basolateral membrane prior to its final appearance in the apical surface (see section III,C). A certain amount of $Na^+K^+ATPase$ is found in association with the apical brush border of kidney tubular cells by immunocytochemistry [143, 144]. A variable amount of VSV G protein is always detected in the apical surface of infected MDCK cells. Given the intense cytopathic effect of VSV, which causes a rather early disruption of the tight junctions [193], the meaning of this finding is unclear at present. The demonstration of mechanisms to translocate proteins from one epithelial surface to another (such as the receptor-mediated transepithelial transport of immunoglobulins; see section II,D) suggests that specific retrieval of surface proteins may be an important factor in sorting. Experiments following the fate of exogenously added contralateral proteins (for example, via liposome fusion) should be helpful in analyzing this point.

Finally, it is not unconceivable that both mechanisms coexist in epithelial cells. The delivery might be mostly polarized, with specific retrieval of proteins inserted in the wrong surface contributing to a more stringent segregation.

IV. SUMMARY AND FUTURE DIRECTIONS IN THE FIELD

The highly polarized nature of the epithelial cell is reflected both in its cytoplasmic and surface organizations. Intracellularly, organelles and cytoskeleton are distributed asymmetrically. The nucleus is usually in the basal one third; the Golgi apparatus, secretory vacuoles, and lysosomes in the apical two thirds. Actin microfilaments in combination with several associated proteins form highly organized structures in the apical brush border of some epithelia. Intermediate filaments, formed by tissue-specific desmins, run across the cell interconnecting spot desmosomes and contributing to determine cell shape. Microtubules, concentrated on the apical cytoplasm and around the Golgi apparatus, possibly play a role in exocytosis. The surface membrane exhibits a striking asymmetric distribution of proteins, lipids, and carbohydrates, responsible for the vectorial function. The apical surface carries out mostly specific functions of each epithelium. Secretory vesicles in exocrine cells fuse only with this region of the plasma membrane. The basolateral surface, like the plasma membrane of nonepithelial cells, relates to the internal medium, receives hormonal signals, and carries out the metabolic exchanges necessary for cell survival. In addition, it bears receptors for attachment to the basal lamina (in which secretion it participates), and forms junctional elements with other cells. Tight junctions or zonulae occludentes limit, to some extent, the diffusion of molecules between the two domains, and seal the intercellular space to the passage of macromolecules. Desmosomes bring strength to cell-cell interactions; gap junctions participate in cell-cell communication.

Epithelial cell lines, which preserve in culture polarity properties of natural epithelia, have become in recent years useful tools to study epithelial polarity and tight-junction development. Their infection with enveloped RNA viruses results in polarized budding from the apical or basolateral surface, depending on the virus. Myxo-, paramyxo-, and some retroviruses bud from the apical surface, and vesicular stomatitis virus is assembled from basolateral membranes. The viral envelope glycoproteins are probably the main determinants of this polarity since they are preferentially segregated into the budding surface prior to assembly; this segregation is stricter for apical viruses than for vsv, suggesting that other factors such as the peripheral protein M may also play a role. The carbohydrates of viral glycoproteins are not determinants of virus-budding polarity. Dissociation of epithelial cells results in disappearance of surface polarity, and consequently in random viral budding. However, polarized viral budding is observed in isolated epithelial cells attached to a substrate, even in the absence of complete cell-cell interactions and tight junctions. Monolayers of epithelial cells double-infected with apical and basolateral viruses preserve the correct virus-budding polarity. Double-labeling immunoelectron microscopic localization experiments on these monolayers reveal the presence of both types of glycoproteins in the Golgi apparatus, indicating that their sorting occurs during or after their passage through this organelle. Primary cultures of thyroid cells exhibit the ability to reverse their intracellular and surface polarity depending on the culture conditions. This property does not, however, appear to be generally shared by other epithelial cells.

It is reasonable to expect that research by workers in the field will continue the exploration of some basic questions which, in spite of recent advances, remain unanswered:

What determines the intracellular and surface structural polarization of epithelial cells? This question relates to the general problem of how genetic information is transduced into a particular cytoarchitecture and shape in eukaryotic cells [233, 234]. The considerable increase in our knowledge of the structure of the apical brush border of intestinal cells in recent years provides an example of the fruitfulness of a combination of immunochemical and cell biological approaches. An important, elusive problem, remains the characterization of the tight junctional protein(s). Monoclonal antibody technology in conjunction with immunocytochemistry could be useful instruments in its identification.

How stringent is the asymmetry in the distribution of surface components in epithelial cells? We need more quantitative data on the degree of polarization of different surface components. Whereas some proteins seem to be almost completely restricted to one plasmalemma domain, others (eg, IgA, IgG receptors) appear to be present on both regions. Why? A related question is, Are lipids polarized? Because the lipid composition of enveloped RNA viruses reflect the original plasmalemmal composition, they should constitute an invaluable help to study this point. What is the polarity of carbohydrate moieties? Immunocytological procedures with lectins could provide very useful hints. Further refinements in the isolation of basolateral and apical membranes are necessary to give definite answers to several of these questions.

What are the sorting-out signals responsible for segregation of apical and basolateral proteins? A genetic approach is probably indicated here. The model of virus-infected epithelial monolayers should be very helpful in this area, since it allows for the search of viral mutants with opposite budding polarity. If the information for sorting out of apical and basolateral proteins is so different that cannot be exchanged by genetic mutation, test-tube "mutants" (or chimeric genes) manufactured by recombinant DNA technology, in combination with technology to introduce and express foreign genes in eukaryotic cells, provide a viable alternative. These approaches will probably yield information on the localization in the protein (and possibly even the sequence) of the regions containing information for sorting out. This will be far from elucidating the mechanism, though, as it can be seen from the difficulties in understanding *how* the signal sequence mediates the vectorial transfer of polypeptides across the ER membrane.

What are the cellular structures involved in the segregation of epithelial surface proteins? Immunocytological techniques at the electron-microscopic level have to be improved both in resolution and sensitivity in order to allow the detection of possible vesicular structures participating in the transport of proteins from Golgi to the plasma membrane. Techniques need to be developed to follow the intracellular migration of individual proteins. An approach based on the introduction via liposomes of proteins in the "wrong" surface (for example, basolateral proteins into the apical surface) may provide interesting clues on the mechanisms of segregation postinsertion in the plasma membrane.

Can the pathways of apical and basolateral proteins be pharmacologically or genetically dissociated? The finding that ionophores affect the intracellular migration of secretory and viral envelope glycoproteins [52, 235], and the observation that monensin may selectively block the intracellular migration of basal but not apically directed viral glycoproteins in epithelial cells [236], constitute interesting avenues for exploration. Finally, if different cellular structures are involved in segregation of epithelial surface proteins, cell mutants could be possibly obtained with defects that selectively affect one of the two pathways.

ACKNOWLEDGMENTS

I am grateful to Drs John A Lewis, George K Ojakian, and Enzo Bard for reading and discussing the manuscript. I thank Drs George K Ojakian and Eva Cramer for providing some of their electronmicrographs. This work was supported by a grant from the National Science Foundation (PCM 8011053A02) and by a Hirschl award.

V. REFERENCES

1. Palade GE: Intracellular aspects of the process of protein secretion. Science 189:347–358, 1975.

2. Sabatini DD, Kreibich G, Morimoto T, Adesnik M: Mechanisms for the incorporation of proteins in membranes and organelles. J Cell Biol 92:1—22, 1982.
3. Rothman JE, Lenard J: Membrane asymmetry. The nature of membrane asymmetry provides clues to the puzzle of how membranes are assembled. Science 194:743–753, 1977.
4. Lodish HF, Braell WA, Schwartz AL, Strous GJAM, Zilberstein A: Synthesis and assembly of membrane and organelle proteins. Int Rev Cytol 12:247–307, 1980.
5. Blobel G: Intracellular protein topogenesis. Proc Natl Acad Aci USA 77:1496–1500, 1980.
6. Chua N-H, Schmidt GW: Transport of proteins into mitochondria and chloroplasts. J Cell Biol 81:461–483, 1979.
7. Jamieson JD: Transport and discharge of exportable proteins in pancreatic acinar cells: In vitro studies. Curr Top Membr Transport 3:273–338, 1972.
8. Jamieson JD: Processing of exportable proteins: Structure function correlates and role of cellular membranes. In Andreoli TH, Hoffman JF, Fanestil DD (eds): "Physiology of Membrane Disorders." New York and London: Plenum Press, 1978, pp 447–458.
9. Sly W, Stahl P: Receptor mediated uptake of lysosomal enzymes. In Silverstein S (ed): "Transport of Macromolecules in Cellular Systems. Life Science Research Report II." Berlin: Dahlem Konferenzem, 1978, pp 229–245.
10. Novikoff AB: The endoplasmic reticulum: cytochemist's view. A review. Proc Natl Acad Sci USA 73:2781–2787, 1976.
11. Lazarow PB: Functions and biogenesis of peroxisomes. In Schweiger HG (ed): "International Cell Biology 1980–1981." Berlin: Springer-Verlag, 1980, p 633.
12. Morre J: Membrane biogenesis. Ann Rev Plant Physiol 26:441–481, 1975.
13. Parodi AJ, Leloir LF: The role of lipid intermediates in the glycosylation of proteins in the eukaryotic cell. Biochim Biophys Acta 559:1–37, 1979.
14. Lennarz WJ: "The Biochemistry of Glycoproteins and Proteoglycans." New York: Plenum Press, 1981, pp 35–84.
15. Watson JD: "Molecular Biology of the Gene." Menlo Park, California: Benjamin Inc, 1970.
16. Paine PL, Horowitz SB: The movement of material between nucleus and cytoplasm. In Prescott DM, Goldstein L (eds): "Cell Biology. A Comprehensive Treatise," 4:299–339, 1980.
17. Snider MD, Sultzman LA, Robbins PW: Transmembrane location of oligosaccharide-lipid synthesis in microsomal vesicles. Cell 21:385–392, 1980.
18. Devillers-Thiery A, Kindt T, Scheele G, Blobel G: Homology in amino terminal sequence of precursors to pancreatic secretory proteins. Proc Natl Acad Sci USA 72:5016–5020, 1975.
19. Lingappa VR, Katz FN, Lodish HF, Blobel G: A signal sequence for the insertion of a transmembrane glycoprotein. Similarities to the signals of secretory proteins in primary structure and function. J Biol Chem 253:8667–8670, 1978.
20. Dobberstein B, Blobel G, Chua N-H: In vitro synthesis and processing of a putative precursor for the small subunit of ribulose-1,5-biphosphate carboxylase of chlamydomonas reinhardii. Proc Natl Acad Sci USA 74:1082–1085, 1977.
21. Maccecchini ML, Rudin Y, Schatz G: Transport of proteins across the mitochondrial outer membrane. J Biol Chem 254:7468–7471, 1979.
22. Maccecchini ML, Rudin Y, Blobel G, Schatz G: Import of proteins into mitochondria: Precursor forms of the extramitochondrially made F_1-ATPase subunits in yeast. Proc Natl Acad Sci USA 76:343–347, 1979.
23. Blobel G, Water P, Chang CN, Goldman B, Erickson AH, Lingappa VR: Translocation of proteins across membranes: the signal hypothesis and beyond. In Hopkins CR and Duncan CJ (eds): "Symposium of the Society of Experimental Biology (Great Britain)." London: Cambridge University Press, 33:9, 1979.

160   **Rodriguez-Boulan**

24. Lingappa VR, Shields D. Woo SLC, Blobel G: Nascent chicken ovalbumin contains the functional equivalent of a signal sequence. J Cell Biol 79:567–572, 1978.
25. Bar-Nun S, Kreibich G, Adesnik M, Alterman L, Negishi M, Sabatini DD: Synthesis and insertion of cytochrome P_{450} into endoplasmic reticulum membranes. Proc Natl Acad Sci USA 77:965–969, 1980.
26. Sabban E, Marchesi V, Adesnik M, Sabatini DD: Erythrocyte membrane protein band 3: Its biosynthesis and incorporation into membranes. J Cell Biol 91:637–646, 1981.
26a. Braell WA, Lodish HF: The erythrocyte anion transport protein is cotranslationally inserted into microsomes. Cell 28:23–31, 1982.
27. Goldman BM, Blobel G: Biogenesis of peroxysomes: Intracellular site of synthesis of catalase and uricase. Proc Natl Acad Sci USA 75:5066–5070, 1978.
28. Neufeld EF, Sando GN, Garvin AJ, Rome LH: The transport of lysosomal enzymes. J Supramol Struct 6:95–101, 1977.
29. Blobel G, Walter P, Chang CN, Goldman BM, Erickson AH, Lingappa VR: Translocation of proteins across membranes: The signal hypothesis and beyond. Symp Soc Exp Biol 33:9–36, 1979.
30. Davis BD, Tai PC: The mechanism of protein secretion across membranes. Nature 283:433–438, 1980.
31. Emr SD, Hall MN, Silhavy TJ: A mechanism of protein localization: The signal hypothesis and bacteria. J Cell Biol 86:701–711, 1980.
32. Lenard J, Compans RW: The membrane structure of lipid-containing viruses. Biochim Biophys Acta 344:51–94, 1974.
33. Lenard J: Virus envelopes and plasma membranes. Annu Rev Biophys Bioeng 7:139–165, 1978.
34. Meldolesi J, Borgese N. De Camilli P, Ceccarelli B: Cytoplasmic membranes and the secretory process. In Poste G, Nicolson (eds): "Membrane Fusion." Amsterdam: Elsevier/ North-Holland, 1978, pp 509–627.
35. Siekevitz P, Palade G: A cytochemical study of the pancreas of the guinea pig. J Biophys Biochem Cytol 7:630–644, 1960.
36. Redman CM, Siekevitz P, Palade GE: Synthesis and transfer of amylase in pigeon pancreatic microsomes. J Biol Chem 241:1150–1158, 1966.
37. Redman CM, Sabatini DD: Vectorial discharge of peptides released by puromycin from attached ribosomes. Proc Natl Acad Sci USA 56:608–615, 1966.
38. Blobel G, Dobberstein B: Transfer of proteins across membranes. I. Presence of proteolytically processed and unprocessed nascent immunoglobulin light chains on membrane-bound ribosomes of murine myeloma. J Cell Biol 67:835–851, 1975.
39. Blobel G, Dobberstein B: Transfer of proteins across membranes. II. Reconstitution of functional rough microsomes from heterologous components. J Cell Biol 67:852–862, 1975.
40. Sabatini DD, Tashiro Y, Palade GE: On the attachment of ribosomes to microsomal membranes. J Mol Biol 19:503–524, 1966.
41. Adelman MR, Sabatini DD, Blobel G: Ribosome-membrane interactions: Nondestructive diassembly of rat liver rough microsomes into ribosomal and membranous components. J Cell Biol 56:206–229, 1973.
42. Borgese N, Mock W, Kreibich G, Sabatini DD: Ribosomal-membrane interaction. In vitro binding of ribosomes to microsomal membranes. J Mol Biol 88:559–580, 1974.
43. Kreibich G, Czako-Graham M, Grebenau R, Mok W, Rodriguez-Boulan E, Sabatini DD: Characterization of the ribosomal binding site in rat liver rough microsomes: Ribophorins I and II, two integral membrane proteins related to ribosome binding. J Supramol Struct 8:279–302, 1978.

44. Sabatini DD, Kreibich G: Functional specialization of membrane-bound ribosomes in eukaryotic cells. In Martonosi A (ed): "The Enzymes of Biological Membranes," Vol 2. New York: Plenum, 1976, pp 531–579.
45. Blobel G, Sabatini DD: Ribosome-membrane interaction in eukaryotic cells. In Manson LA (ed): "Biomembranes." New York: Plenum, 2:193–195, 1971.
46. Milstein C, Brownlee GG, Harrison TM, Mathews MB: A possible precursor of immunoglobulin light chains. Nature New Biol 239:117–120, 1972.
47. Schechter I: Biologically and chemically pure mRNA coding for mouse immunoglobulin L-chain prepared with the aid of antibodies and immobilized oligothymidine. Proc Natl Acad Sci USA 70:2256–2260, 1973.
48. Kemper B, Habener JF, Mulligan RC, Potts JT Jr, Rich A: Preproparathyroid hormone: A direct translation product of a parathyroid messenger RNA. Proc Natl Acad Sci USA 71:3731–3735, 1974.
49. Swan D, Aviv H, Leder P: Purification and properties of biologically active messenger RNA for a myeloma light chain. Proc Natl Acad Sci USA 69:1967–1971, 1972.
50. Anfinsen CB: The formation of the tertiary structure of proteins. Harvey Lect 61:95–116, 1966.
51. Olsen BR, Berg RA, Kishida Y, Prockop DJ: Further characterization of embryonic tendon fibroblasts and the use of immunoferritin techniques to study collagen biosynthesis. J Cell Biol 43:289–311, 1975.
52. Tartakoff AM, Vassalli P: Plasma cell immunoglobulin secretion: Arrest is accompanied by alterations of the Golgi complex. J Exp Med 146:1332–1345, 1978.
53. Rose JK: Complete intergenic and flanking sequences from the genome of vesicular stomatitis virus. Cell 19:415–421, 1980.
54. Garoff H: The spike glycoprotein of semliki forest virus A model for plasma membrane proteins. In Schweiger HG (ed): "International Cell Biology 1980–1981." Berlin: Springer-Verlag, 1981, p 572.
55. Garoff H, Frischauf AM, Simmons K, Lehrach H, Delius H: Nucleotide sequence of cDNA coding for semliki forest virus glycoproteins. Nature 288:236–241, 1980.
56. Minjou W, Threlfall G, Verhoeyen M, Devos R, Saman E, Fang R, Huylebroek D, Fiers W, Barber C, Carey N, Emtage S: Complete structure of hemagglutinin gene from the human influenza A/Victoria/3/75 (H3N2) strain as determined from cloned DNA. Cell 19:683–696, 1980.
57. Katz FN, Rothman JE, Lingappa VR, Blobel G, Lodish HF: Membrane assembly in vitro: Synthesis, glycosylation and asymetric insertion of a transmembrane protein. Proc Natl Acad Sci USA 74:3278–3282, 1977.
58. Morrison TG, Lodish HF: Site of synthesis of membrane and nonmembrane proteins of vesicular stomatitis virus. J Cell Biol 250:6955–6962, 1975.
59. Atkinson PH: Glycoprotein and protein precurssor to plasma membranes in VSV infected HeLa cells. J Supramol Struct 8:89–109, 1978.
60. Katz FN, Lodish HF: Transmembrane biogenesis of the vesicular stomatitis virus glycoprotein. J Cell Biol 80:416–426, 1979.
61. Rose JK, Welch WJ, Sefton BM, Esch FS, Ling NC: Vesicular stomatitis virus glycoprotein is anchored in the viral membrane by a hydrophobic domain near the COOH terminus. Proc Natl Acad Sci USA 77:3884–3888, 1980.
62. Robertson MA, Etchison JR, Robertson JS, Summers DF, Stanley P: Specific changes in the oligosaccharide moieties of vesicular stomatitis virus grown in different lectin-resistant CHO cells. Cell 13:515–526, 1978.
63. Tabas I, Schlesinger S, Kornfeld S: Processing of high mannose oligosaccharides to form complex type oligosaccharides on the newly synthesized polypeptides of the vesicular stomatitis virus G protein and the IgG heavy chain. J Biol Chem 253:716–722, 1978.

64. Hunt LA, Etchison JR, Summers DF: Oligosaccharide chains are trimmed during synthesis of the envelope glycoproteins of vesicular stomatitis virus. Proc Natl Acad Sci USA 75:754–758, 1978.

65. Bergman JE, Tokoyasu KT, Singer SJ: Passage of an integral membrane protein, the vesicular stomatitis virus glycoprotein, through the Golgi apparatus en route to the plasma membrane. Proc Natl Acad Sci USA 78:1746–1750, 1981.

65a. Green J, Griffiths G, Louvard D, Quinn P, Warren G: Passage of viral membrane proteins through the Golgi complex. J Mol Biol 152:663–698, 1981.

66. Rothman JE, Fine RE: Coated vesicles transport newly synthesized membrane glycoproteins from endoplasmic reticulum to plasma membrane in two successive stages. Proc Natl Acad Sci USA 77:780–784, 1980.

67. Rothman JE, Bursztyn-Pettegrew H, Fine RE: Transport of the membrane glycoprotein of vesicular stomatitis virus to the cell surface in two stages by clathrin-coated vesicles. J Cell Biol 86:162–171, 1980.

68. Lazarowitz SG, Goldberg AR, Choppin PW: Proteolytic cleavage by plasmin of the HA polypeptide of influenza virus: Host cell activation of serum plasminogen. Virology 56:172–180, 1973.

69. Nakamura K, Compans RW: The cellular site of sulfation of influenza viral glycoproteins. Virology 79:381–392, 1977.

70. Schulman JL, Palese P: Virulence factors of influenza A viruses: WSN virus neuraminidase required for plaque production in MDBK cells. J Virol 24:170–176, 1977.

71. Schmidt MFG, Schlessinger MJ: Fatty acid binding to vesicular stomatitis virus glycoprotein: A new type of post translational modification of the viral glycoprotein. Cell 17:813–819, 1979.

72. Dobberstein B, Garoff H, Warren G, Robinson PJ: Cell-free synthesis and membrane insertion of mouse H-2D^d histocompatibility antigen and β2-microglobulin. Cell 17:759–769, 1979.

73. Krangel MS, Ohr HT, Stromingger JL: Assembly and maturation of HLA-A and HLA-B antigens in vivo. Cell 18:979–991, 1979.

74. Singer SJ, Nicholson GL: The fluid mosaic model of the structure of cell membranes. Science 175:720–731, 1972.

75. Kreibich G, Ulrich BL, Sabatini DD: Proteins of rough microsomal membranes related to ribosome binding. I. Identification of ribophorins I and II, membrane proteins characteristic of rough microsomes. J Cell Biol 77:464–487, 1978.

76. Louvard D, Reggio H, Warren G: Antibodies to the Golgi complex and the rough endoplasmic reticulum. J Cell Biol 92:92–107, 1982.

76a. Lin JJ-C, Queally SA: A monoclonal antibody that recognizes Golgi-associated protein of cultured fibroblast cells. J Cell Biol 92:108–112, 1982.

77. Farquhar MG: Traffic of products and membranes through the Golgi complex. In Silverstein SC (ed): "Life Sciences Research Report II. Dahlem Konferenzen. Transport of macromolecules in cellular systems," 1978, pp 341–362.

78. Tartakoff AM: The Golgi complex: Crossroads for vesicular traffic. Int Rev Exp Pathol 22:227–251, 1980.

79. Rothman JE: The Golgi apparatus: Two organelles in tandem. Science 213:1212–1219, 1981.

80. Beams H, Kessel R: The Golgi apparatus: Structure and function. Int Rev Cytol 23:209–276, 1973.

81. Whaley W, Dauwalder M: The Golgi apparatus, the plasma membrane and functional integration. Int Rev Cytol 58:199–245, 1979.

82. Morre DJ: The Golgi apparatus and membrane biogenesis. In Poste G, Nicolson GL (eds): "Cell Surface Reviews." Amsterdam: Elsevier/North-Holland, 1977, pp 1–83.

83. Novikoff AB, Novikoff PM: Cytochemical contribution to differentiating GERL from Golgi apparatus. Histochem J 9:525–552, 1977.
84. Bainton DF, Nichols BA, Farquhar MG: Primary lysosomes of blood leukocytes. In Dingle JT, Dean RT (eds): "Lysosomes in Biology and Pathology." Amsterdam: North-Holland, 1976, Vol 5, pp 3–32.
85. Elovson J: Biogenesis of plasma membrane glycoproteins. Tracer kinetic study of two rat liver plasma membrane glycoproteins in vivo. J Biol Chem 255:5816–5825, 1980.
86. Cohn Z, Fedorko, Hirsch J: The in vitro differentiation of mononuclear phagocytes. V. The formation of macrophage lysosomes. J Exp Med 123:757–766, 1966.
87. Strous GJAM, Lodish HF: Intracellular transport of secretory and membrane proteins in hepatoma cells infected by vesicular stomatitis virus. Cell 22:709–718, 1980.
88. Siekevitz P: Biological membranes: The dynamics of their organization. Ann Rev Physiol 34:117–140, 1972.
89. Schimke RT: The synthesis and degradation of membrane proteins. In Ceccarelli B, Clementi F, Meldolesi J (eds): "Advances in Cytopharmacology." 1975, Vol 2, pp 63–69.
90. Tweto J, Doyle D: Turnover of proteins of the eukaryotic cell surface. In Poste G, Nicolson GL (eds): "Cell Surface Reviews." Amsterdam: Elsevier/North-Holland, 1977, Vol 4, pp 137–164.
91. Morre DJ, Kartenbeck J, Franke WW: Membrane flow and interconversions among endomembranes. Biochim Biophys Acta 559:71–152, 1979.
92. Castle JD, Jamieson JD, Palade GE: Secretion granules of the rabbit parotid gland. Isolation, subfractionation and characterization of the membrane and content subfractions. J Cell Biol 64:182–210, 1975.
93. Meldolesi J. De Camilli P, Peluchetti D: The membrane of secretory granules: Structure. In Thorn N, Petersen OH (eds): "Secretory Mechanisms in Exocrine Glands." Copenhagen: Munksgaard, 1974, pp 137–151.
94. Silverstein SC, Steinman RM, Cohn ZA: Endocytosis. Ann Rev Biochem 46:669–772, 1977.
95. Marsh M, Helenius A: Adsorptive endocytosis of semliki forest virus. J Mol Biol 142:439–454, 1980.
95a. Steinman RM, Scott EB, Cohn ZA: Membrane flow during pinocytosis. A stereological analysis. J Cell Biol 68:665–687, 1976.
96. Herzog V, Farquhar MG: Uptake of horseradish peroxidase in the Golgi complex of stimulated acinar cells of the parotid gland. Proc Natl Acad Sci USA 74:5073–5079, 1977.
97. Herzog V, Miller F: Membrane retrieval in epithelial cells of isolated thyroid follicles. Eur J Cell Biol 19:203–215, 1979.
98. Herzog V, Reggio H: Pathways of endocytosis from luminal plasma membranes in rat exocrine pancreas. Eur J Cell Biol 21:141–150, 1980.
99. Brown MS, Goldstein JL: Receptor mediated endocytosis: Insights from the lipoprotein receptor system. Proc Natl Acad Sci USA 76:3330–3337, 1979.
100. Pastan IH, Willingham MC: Receptor mediated endocytosis of hormones in cultured cells. Ann Rev Physiol 43:239–250, 1981.
101. Pastan IH, Willingham MC: Journey to the center of the cell: Role of the receptosome. Science 214:504–509.
102. Rodewald RB: Selective antibody transport in the proximal small intestine of the neonatal rat. J Cell Biol 45:635–640, 1970.
103. Brandtzaeg P, Baklien K: Intestinal secretion of IgA and IgM: A hypothetical model. In Porter R, Knight J (eds): "Immunology of the Gut." Amsterdam: Elsevier, 1977, pp 77–108.
104. Kraehenbuhl JP, Kuhn L: Transport of immunoglobins across epithelia. In Silverstein SC (ed): "Transport of Macromolecules in Cellular Systems." Berlin: Dahlem Konferenzen, 1978, pp 213–228.

105. Mostov KE, Kraehenbuhl JP, Blobel G: Receptor-mediated transcellular transport of immunoglobin: synthesis of secretory component as multiple and larger transmembrane forms. Proc Nat Acad Sci USA 77:7257–7261, 1980.

106. Brambell FWR: "The Transmission of Passive Immunity From Mother to Young." Amsterdam: North-Holland, 1970.

107. Rodewald RB: Intestinal transport of antibodies in the newborn rat. J Cell Biol 58:189–211, 1973.

108. Rodewald RB: pH-dependent binding of immunoglobins to intestinal cells of the neonatal rat. J Cell Biol 71:666–670, 1976.

109. Jones EA, Waldman TA: The mechanism of intestinal uptake and transcellular transport of IgG in the neonatal rat. J Clin Invest 51:2916–2927, 1972.

110. Palade GE, Simionescu M, Simionescu N: Transport of solutes across the vascular endothelium. In Silverstein SC (ed): "Transport of Molecules in Cellular Systems." Berlin: Dahlem Konferenzen, 1978, pp 145–166.

111. Maunsbach AB, Christensen EI, Ottosen PD, Larsson L: Uptake, transport, and digestion of proteins in kidney cells. In Takeuchi T, Ogawa K, Fujita S (eds): "Proc IV Intern Congr Histochem Cytochem." Kyoto: Nakamishi, 1972, p 65.

112. Van Devrs B, Moller M, Amtorp O: Uptake of horseradish peroxidase from CSF into the choroid plexus of the rat, with special reference to transepithelial transport. Cell Tiss Res 187:215–234, 1978.

113. Abrahamson DR, Rodewald RB: Evidence for the sorting of endocytic vesicle contents during the receptor mediated transport of IgG across the newborn rat intestine. J Cell Biol 91:270–280, 1981.

114. Berridge MJ, Oschman JL: "Transporting Epithelia." New York and London: Academic Press, 1972.

115. Brunser O, Luft JH: Fine structure of the apex of absorptive cells from rat small intestine. J Ultrastruct Res 31:391, 1970.

116. Mukherjee TM, Staehelin LA: The fine structural organization of the brush border of intestinal epithelial cells. J Cell Sci 8:573–599, 1971.

117. Mooseker MS, Tilney LG: Organization of an actin filament membrane complex. Filament polarity and membrane attachment in the microvilli of intestinal epithelial cells. J Cell Biol 67:725–743, 1975.

118. Mooseker MS: Actin filament-membrane attachment in the microvilli of intestinal epithelial cells. In Goldman R, Pollard T, Rosenbaum J (eds): "Cell Motility." Cold Spring Harbor, New York: Cold Spring Harbor Lab, 1976, pp 631–650.

119. Hull BE, Staehelin LA: The terminal web: A reevaluation of its structure and function. J Cell Biol 81:67–82, 1979.

120. Bretscher AP, Weber K: Purification of microvilli and analysis of the protein components of the microfilament core bundle. Exp Cell Res 116:397–407, 1978.

121. Bretscher AP, Weber K: Localization of actin and microfilament-associated proteins in the microvilli and terminal web of the intestinal brush border by immunofluorescence microscopy. J Cell Biol 79:839–845, 1978.

122. Bretscher AP, Weber K: Villin: The major microfilament-associated protein present in microvilli and other cell surface structures. Proc Natl Acad Sci USA 76:2321–2325, 1979.

123. Bretscher AP, Weber K: Fimbrin, a new microfilament-associated protein present in microvilli and other cell surface structures. J Cell Biol 86:335–340, 1980.

124. Craig SW, Pardo JV: Alpha actinin localization in the junctional complex of intestinal epithelial cells. J Cell Biol 80:203–210, 1979.

125. Geiger B, Tokuyasu KT, Singer SJ: Immunocytochemical localization of α-actinin in intestinal epithelial cells. Proc Natl Acad Sci USA 76:2833–2837, 1979.

126. Geiger B, Dutton AH, Tokuyasu KT, Singer SJ: Immunoelectron microscope studies of membrane-microfilament interactions: Distributions of α-actinin, tropomyosin and vinculin in intestinal epithelial brush border and chicken gizzard smooth muscle cells. J Cell Biol 91:614–628, 1981.

127. Geiger B, Tokuyasu KT, Dutton AH, Singer SJ: Vinculin, an intracellular protein localized at specialized sites where microfilament bundles terminate at cell membranes. Proc Natl Acad Sci USA 77:4217–4131, 1980.

128. Lazarides E, Burridge K: α-Actinin: Immunofluorescence localization of a muscle structural protein in non muscle cells. Cell 6:289–298, 1975.

129. Geiger B: A 130K protein from chicken gizzard. Its localization at the termini of microfilament bundles in cultured chicken cells. Cell 18:193–205, 1979.

130. Winter S, Jarasch ED, Schmid E, Franke WW, Denk H: Differences in polypeptide composition of cytokeratin filaments, including tonofilaments, from different epithelial tissues and cells. Eur J Cell Biol 22:371, 1980.

131. Soifer D: The biology of cytoplasmic microtubules. Ann NY Acad Sci 253, 1975.

132. Muller J. Kachadorian WA, Discala VA: Evidence that ADH stimulated intramembrane particle aggregates are transferred from cytoplasmic to luminal membranes in toad bladder epithelial cells. J Cell Biol 85:83–95, 1980.

133. Chevalier J, Bourguet J, Hugon JS: Membrane associated particles. Distribution in frog urinary bladder epithelium at rest and after oxytocin treatment. Cell Tissue Res 152:129–140, 1974.

134. Humbert F, Montesano R, Grosso A, DeSouza RC, Orci L: Particle aggregates in plasma and intracellular membranes of toad bladder (granular cell). Experientia 33:1364–1367, 1977.

135. Wade JB: Membrane structural specialization of the toad urinary bladder revealed by freeze fracture technique. III. Localization, structure and vasopressin dependence of intramembranous particle arrays. J Membrane Biol Sp Issue 281–296, 1978.

136. Lojda Z: Cytochemistry of enterocytes and of other cells in the mucous membrane of the small intestine. In Smith DH (ed): "Intestinal Absorption," Biomembranes, Vol 4A. London: Plenum, 1974, pp 43–122.

137. Gomori G: The distribution of phosphatase in normal organs and tissues. J Cell Comp Physiol 17:71–83, 1941.

138. Desnuelle P: Intestinal and renal aminopeptidase: A model of a transmembrane protein. Eur J Biochem 101:1–11, 1979.

139. Semenza G: Small intestinal disaccharides: Their properties and role as sugar translocators across natural and artificial membranes. In Martonosi A (ed): "Membrane transport," Enzymes of Biological Membranes, Vol 3. New York: Plenum, 1976, pp 349–382.

140. Schwartz IL, Shlatz LJ, Kinne-Safran E, Kinne R: Target cell polarity and membrane phosphorylation in relation to the mechanism of action of antidiuretic hormone. Proc Natl Acad Sci USA 71:2595–2599, 1974.

141. Murer H, Kinne R: The use of isolated membrane vesicles to study epithelial transport processes. J Memb Biol 55:81–95, 1980.

142. Stirling CR: Radioautographic localization of sodium pump sites in rabbit intestine. J Cell Biol 53:704–714, 1972.

143. Kyte J: Immunoferritin determination of the distribution of Na^+K^+ATPase over the plasma membranes of the renal convoluted tubules. 1. Distal segment. J Cell Biol 68:287–303, 1976.

144. Kyte J: Immunoferritin determination of Na^+K^+ATPase over the plasma membrane of renal convoluted tubules. II. Proximal segment. J Cell Biol 68:304–318, 1976.

145. Ernst SA, Mills JW: Basolateral plasma membrane localization of ouabain-sensitive sodium transport sites in the secretory epithelium of the avian salt gland. J Cell Biol 75:74–94, 1977.

146. Kirby WN, Parr EL: The occurrence and distribution of H_2 antigens on mouse intestinal epithelial cells. J Histochem Cytochem 27:746–750, 1979.

147. Wall DA, Hubbard AL: Galactose specific recognition system of mammalian liver: Receptor distribution on the hepatocyte surface. J Cell Biol 90:687–696, 1981.

148. Rodewald R: Distribution of immunoglobulin G receptors in the small intestine of the young rat. J Cell Biol 85:18–32, 1980.

149. Bentley PJ: The physiology of the urinary bladder of amphibia. Biol Rev 41:275–316, 1966.

150. Sugrue SP, Hay ED: Response of basal epithelial cell surface and cytoskeleton to solubilized extracellular matrix molecules. J Cell Biol 91:45–54, 1981.

151. Kleinman HK, Klebe RJ, Martin GR: Role of collagenous matrices in the adhesion and growth of cells. J Cell Biol 88:473–485, 1981.

152. Maylie-Pfenninger MF, Jamieson JD: Distribution of cell surface saccharides on pancreatic cells. J Cell Biol 80:77–95, 1979.

153. Muresan V, Jamieson JD: Asymmetric distribution of sialo-glycoconjugates on the plasmalemma of pancreatic acinar cells. Eur J Cell Biol 22:276, 1980.

154. Forstner GG, Tanaka K, Isselbacher KJ: Lipid composition of the isolated rat intestinal microvillus membrane. Biochem J 109:51–59, 1968.

154a. Douglas AP, Kerley R, Isselbacher KJ: Preparation and characterization of the lateral and basal plasma membranes of the rat intestinal epithelial cell. Biochem J 128:1329–1338, 1972.

155. Kawai K, Fujita M, Nakao M: Lipid components of two different regions of an intestinal epithelial cell membrane of mouse. Biochim Biophys Acta 369:222–233, 1974.

155a. Lewis BA, Gray GM, Coleman R, Michell RH: Differences in the enzymic, polypeptide, glycopeptide, glycolipid and phospholipid compositions of plasma membranes from the two surfaces of intestinal epithelial cells. Biochem Soc Transact 3:752–753, 1975.

156. Brasitus TA, Schachter D: Lipid dynamics and lipid-protein interactions in rat enterocyte basolateral and microvillus membranes. Biochemistry 19:2763–2769, 1980.

156a. Dragsten PR, Blumenthal R, Handler JS: Membrane asymmetry in epithelia: Is the tight junction a barrier to diffusion in the plasma membrane? Nature 294:718–722, 1981.

157. Farquhar MG, Palade GE: Junctional complexes in various epithelia. J Cell Biol 17:375–412, 1963.

158. Goodenough DA, Revel JP: A fine structural analysis of intracellular junctions in the mouse liver. J Cell Biol 45:272–290, 1970.

159. Erlij D, Martinez-Palomo A: Role of tight junctions in epithelial function. In Giebisch G, Tosteson DC, Ussing HH (eds): "Membrane Transport in Biology." New York: Springer-Verlag, 1978, pp 27–53.

160. Cogoli A, Mosiman H, Vock C, Balthazar AK, Semenza G: A simplified procedure for the isolation of the sucrase-isomaltase complex from rabbit intestine. Eur J Biochem 30:7–14, 1972.

161. Louvard D, Maroux S, Desnuelle P: Topological studies on the hydrolases bound to the intestinal brush border membrane. II. Interactions of free and bound aminopeptidase with a specific antibody. Biochim Biophys Acta 389:389–400, 1975.

162. Maroux S, Louvard D: On the hydrophobic part of aminopeptidase and maltases which bind the enzyme to the intestinal brush border membrane. Biochim Biophys Acta 419:189–195, 1976.

163. Brunner J, Hauser H, Braun H, Wilson KJ, Wacker H, O'Neill B, Semenza G: The mode of association of the enzyme complex sucrase-isomaltase with the intestinal brush border membrane. J Biol Chem 254:1821–1828, 1979.

164. Frank G, Brunner J, Hauser H, Wacker H, Semenza G, Zuber H: The hydrophobic anchor of small intestinal sucrase-isomaltase. N-terminal sequence of the isomaltase subunit. FEBS Lett 96:183–188, 1978.

165. Hauri HP, Quaroni A, Isselbacher KJ: Biogenesis of intestinal plasma membrane post-translational route and cleavage of sucraser isomaltase. Proc Natl Acad Sci USA 76:5183–5186, 1979.

166. Reaven EP, Reaven GM: Distribution and content of microtubules in relation to the transport of lipid. An ultrastructural quantitative study of the absorptive cell of the small intestine. J Cell Biol 75:559–572, 1977.

167. Quaroni A, Kirsch K, Weiser MM: Synthesis of membrane glycoproteins in rat small intestinal villus cells. Effect of colchicine on the redistribution of L 1,5,6-^{3}H fucose labelled membrane glycoproteins among Golgi lateral, basal and microvillus membranes. J Biochem 182:213–221, 1979.

168. Kyte J: Properties of the two polypeptides of sodium and potassium dependent adenosine tryphosphatase. J Biol Chem 247:7642–7649, 1972.

169. Craig WS, Kyte J: Stiochiometry and molecular weight of the minimum asymmetric unit of canine renal sodium and potassium-ion activated adenosine triphosphatase. J Biol Chem 255:6262–6269, 1980.

170. Kyte J: Molecular considerations relevant to the mechanism of active transport. Nature 292:201–204, 1981.

171. Kyte J: Structural studies of sodium and potassium ion-activated adenosine triphosphatase. J Biol Chem 250:7443–7449, 1975.

172. Klinenberg M: Membrane protein oligomeric structure and transport function. Nature 290:449–454, 1981.

173. Sherman JM, Sabatini DD, Morimoto T: Studies on the biosynthesis of the polypeptide subunits of the Na$^+$K$^+$ATPase of kidney cells. J Cell Biol 87:307a, 1980.

174. Rabito CA, Tchao R, Valentich J, Leighton J: Distribution and characteristics of the occluding junctions in a monolayer of a cell line (MDCK) derived from canine kidney. J Memb Biol 43:351, 1978.

175. Leighton J. Brada Z, Estes LW, Justh G: Secretory activity and oncogenicity of a cell line derived from canine kidney. Science 163:472–473, 1969.

176. Madin SH, Darby NB Jr: "American Type Culture Collection Catalogue of Strains II." Rockville, Maryland, 1975, p 47.

177. Handler JS, Perkins FM, Johnson JP: Studies of renal function using cell culture techniques. Am J Physiol 238:F_1–F_9, 1980.

178. Misfeldt DS, Hamamoto ST, Pitelka DR: Transepithelial transport in cell culture. Proc Natl Acad Sci USA 73:1212–1216, 1976.

179. Cereijido M, Robbins ES, Dolan WJ, Rotunno CA, Sabatini DD: Polarized monolayers formed by epithelial cells on a permeable and translucent support. J Cell Biol 77:853–880, 1978.

180. Rindler MJ, Chuman LM, Shaffer L, Saier MH: Retention of differentiated properties in an established dog kidney epithelial cell line (MDCK). J Cell Biol 81:635–648, 1979.

181. Herzlinger D, Beasley E, Ojakian GK: The MDCK cell line expresses a cell surface antigen of the kidney distal tubule. J Cell Biol 91:113a, 1981.

182. Taub M, Chuman L, Saier MH, Sato GH: The growth of a kidney epithelial cell line (MDCK) in hormone-supplanted serum free media. Proc Natl Acad Sci USA 76:3338–3342, 1979.

183. Cramer ER, Milks LC, Ojakian GK: Transepithelial migration of human neutrophils: An in vitro model system. Proc Natl Acad Sci USA 77:4069–4073, 1980.

184. Richardson JCW, Simmons NL: Demonstration of protein asymmetries in the plasma membrane of cultured renal (MDCK) epithelial cells by lactoperoxidase mediated iodination. FEBS Lett 105:201–204, 1979.
185. Hoi-Sang U, Saier MH, Ellisman MH: Tight junction formation is closely linked to the polar redistribution of intramembranous particles in aggregating MDCK epithelia. Exp Cell Res 122:384–391, 1979.
186. Cereijido M, Ehrenfeld J, Meza I, Martinez-Palomo A: Structural and functional membrane polarity in cultured monolayers and MDCK cells. J Membr Biol 52:147–159, 1980.
187. Louvard D: Apical membrane aminopeptidase appears at site of cell-cell contact in cultured kidney epithelial cells. Proc Natl Acad Sci USA 77:4132–4136, 1980.
188. Rodriguez-Boulan E, Sabatini DD: Asymmetric budding of viruses in epithelial monolayers: A model system for study of epithelial polarity. Proc Natl Acad Sci USA 75:5071–5075, 1978.
189. Murphy JS, Bang FB: Observations with the electron microscope on cells of the chick chorio-allantoic membrane infected with influenza virus. J Exp Med 95:259–277, 1952.
190. Bang FB: The development of Newcastle Disease virus in cells of the chorioallantoic membrane as studied by thin section. Bull John Hopk Hosp 92:309–322, 1953.
191. Hess RT, Falcon LA: Observations on the interaction of baculovirusus with the plasma membrane. J Gen Virol 36:525–530, 1977.
192. Pickett PB, Pitelka DR, Hamamoto ST, Misfeldt DS: Occluding junctions and cell behavior in primary cultures of normal and neoplastic mammary gland cells. J Cell Biol 66:316–332, 1975.
193. Rodriguez-Boulan EJ, Pendergast M: Polarized distribution of viral envelope glycoproteins in the plasma membrane of infected epithelial cells. Cell 20:45–54, 1980.
194. Gibson R, Leavitt R, Kornfeld S, Schlessinger S: Synthesis and infectivity of vesicular stomatitis virus containing nonglycosylated G protein. Cell 13:671–679, 1978.
195. Gibson R, Schlessinger S, Kornfeld S: The nonglycosylated glycoprotein of vesicular stomatitis virus is temperature sensitive and undergoes intracellular aggregation at elevated temperatures. J Biol Chem 254:3600–3607, 1979.
196. Green R, Meiss HK, Rodriguez-Boulan EJ: Glycosylation does not determine segregation of viral envelope proteins in the plasma membrane of epithelial cells. J Cell Biol 89:230–239, 1981.
197. Meiss HK, Green R, Rodriguez-Boulan EJ: Lectin resistant mutants of polarized epithelial cells. Molec Cell Biol 2:1287–1294, 1982.
198. Takatsuki A, Khono K, Tamura G: Inhibition of biosynthesis of polyisoprenol sugars in chick embryo microsomes by tunicamycin. Agric Biol Chem 39:2089–2091, 1975.
199. Palese P, Tobita K, Ueda M, Compans RW: Characterization of a temperature sensitive mutant of influenza virus defective in neuraminidase. Virology 61:397–410, 1974.
200. Roth MG, Fitzpatrick J, Compans RW: Polarity of influenza and vesicular stomatitis virus maturation in MDCK cells: Lack of requirement for glycosylation of viral glycoproteins. Proc Natl Acad Sci USA 76:6430–6434, 1979.
201. Moreno JH, Diamond JM: Discrimination of monovalent inorganic cations by "tight" junctions of gallbladder epithelium. J Memb Biol 15:277–318, 1974.
202. Pisam M, Ripoche P: Redistribution of surface macromolecules in dissociated epithelial cells. J Cell Biol 71:907–920, 1976.
203. Ziomek CA, Schulman S, Edidin M: Redistribution of membrane proteins in isolated mouse intestinal cells. J Cell Biol 86:849–857, 1980.
204. Meldolesi J, Castiglioni G, Parma R, Nassivera N. De Camilli P: Ca^{++} dependent disassembly and reassembly of occluding junctions in guinea pig pancreatic acinar cells. Effects of drugs. J Cell Biol 79:156–172, 1978.

205. Damsky CH, Knudsen KA, Dorio RJ, Buck CA: Manipulation of cell-cell and cell-substratum interactions in mouse mammary tumor epithelial cells using broad spectrum antisera. J Cell Biol 89:173–184, 1981.
206. Knudsen KA, Rao PE, Damsky CH, Buck CA: Membrane glycoproteins involved in cell substratum adhesion. Proc Natl Acad Sci USA 78:6071–6075, 1981.
207. Rodriguez-Boulan E, Paskiet K, Sabatini DD: Asymmetric budding of enveloped viruses from isolated epithelial cells attached to a collagen substrate. J Cell Biol 91:121a, 1981.
208. Michl J, Pieczonka MM, Unkeless JC, Silverstein SC: Effects of immobilized immune complexes on Fc- and complement-receptor function in resident and thioglycollate-elicited mouse peritoneal macrophages. J Exp Med 150:607–621, 1979.
209. Goldberg B: Binding of soluble type I collagen molecules to the fibroblast plasma membrane. Cell 16:265–275, 1979.
210. Rubin K, Oldberg A, Hook M, Obrink B: Adhesion of rat hepatocytes to collagen. Exp Cell Res 117:165–177, 1978.
211. Rubin K, Hook M, Obrink B, Timple P: Substrate adhesion of rat hepatocytes: Mechanisms of attachment to collagen substrates. Cell 24:463–470, 1981.
212. Rubin K, Johansson I, Petterson I, Ocklind C, Obrink B, Hook M: Attachment of rat hepatocytes to collagen and fibronectin: A study using antibodies directed against cell surface components. Biochem Biophys Res Commun 91:86–94, 1979.
213. Cereijido M, Rotunno CA, Robbins ES, Sabatini DD: Polarized epithelial membranes produced in vitro. In Hoffman JF (ed): "Membrane Transport Processes." New York: Raven Press, 1978, p 433.
214. Meza I, Ibarra G, Sabanero M, Martinez-Palomo A, Cereijido M: Occluding junctions and cytoskeletal components in a cultured transporting epithelium. J Cell Biol 87:746–754, 1980.
215. Griepp EB, Robbins ES, Dolan WJ, Sabatini DD: Role of microfilaments in tight junction formation. J Cell Biol 91:118a, 1981.
216. DiPasquale A: Locomotory activity of epithelial cells in culture. Exp Cell Res 94:191–217, 1975.
217. Marcus P: Dynamics of surface modification in myxovirus-infected cells. Cold Spring Harbor Symp Quant Biol 27:351–365, 1962.
218. Vasiliev JM, Gelfand IM, Domnina LV, Pletyushkina OY: Active cell edge movement of concanavalin A receptors of the surface of epithelial and fibroblastic cells. Proc Natl Acad Sci USA 73:4085–4089, 1976.
219. Vaughan RB, Trinkaus JP: Movements of epithelial sheets in vitro. J Cell Sci 1:407–413, 1966.
220. Chambard M, Gabrion J, Mauchamp J: Influence of collagen gel on the orientation of epithelial cell polarity: follicle formation from isolated thyroid cells and from preformed monolayers. J Cell Biol 91:157–166, 1981.
221. Mauchamp J, Gabrion J, Bernard P: Morphological and electrical polarity of the thyroid epithelium reconstructed in culture. Inserm 85:43–52, 1979.
222. Nitsch L, Wollman SH: Ultrastructure of intermediate stages in polarity reversal of thyroid epithelium in follicles in suspension culture. J Cell Biol 86:875–880, 1980.
223. Gillman J: The cellular cycle, the Golgi apparatus and the phenomenon of reversal in the human thyroid parenchyma. Anat Rec 60:209–230, 1934.
224. Hay ED, Revel JP: Fine structure of the developing avian cornea. In Wolsky A, Chen PS (eds): "Monographs in Developmental Biology." Basel: Kargel, 1969, Vol 1.
225. Trelstad RL: The Golgi apparatus in chick corneal epithelium: Changes in intracellular position during development. J Cell Biol 45:34–42, 1970.

226. Beams HW, King RL: The Golgi apparatus in the developing tooth with special reference to polarity. Anat Rec 57:29, 1933.
227. Rindler MJ, Ivanov IE, Rodriguez-Boulan EJ, Sabatini DD: Biogenesis of epithelial cell plasma membranes. In "Membrane Recycling," CIBA Foundation Symposium, Vol 92, 1982 (in press). Pitman Books Ltd, London.
228. Choppin PW, Compans RW: Phenotypic mixing of envelope proteins of the parainfluenza virus sv5 and vesicularstomatitis virus. J Virol 5:609–616, 1970.
229. McSharry JJ, Compans RW, Choppin PW: Proteins of vesicular stomatitis virus and of phenotypically mixed vesicular stomatitis-simian virus 5 virions. J Virol 8:722–729, 1972.
230. Roth MG, Compans RW: Delayed appearance of pseudotypes between vesicular stomatitis virus and influenza virus during mixed infection of MDCK cells. J Virol 40:848–860, 1981.
231. Ojakian GK: Tumor promoter-induced changes in the permeability of epithelial cell tight junctions. Cell 23:95–103, 1981.
232. Martinez-Palomo A, Meza I, Beaty G, Cereijido M: Experimental modulation of occluding junctions in a cultured transporting epithelium. J Cell Biol 87:736–745, 1980.
233. Solomon F: Specification of cell morphology by endogenous determinants. J Cell Biol 90:547–553, 1981.
234. Fulton AB: How do eukaryotic cells construct their cyto architecture. Cell 24:4–5, 1981.
235. Johnson DC, Schlessinger MJ: Vesicular stomatitis virus and Sindvis virus glycoprotein transport to the cell surface is inhibited by ionophores. Virology 103:407–424, 1980.
236. Alonso FB, Compans RW: Differential effect of monensin on enveloped viruses that form at distinct plasma membrane domains. J Cell Biol 89:700–705, 1981.

Modern Cell Biology, 1:171–197

The Acetylcholine Receptor–Ion Channel Complex: Linkage Between Binding and Response

Alfred Maelicke and Heino Prinz

From the Max-Planck-Institut für Ernährungsphysiologie, Reinlanddamm 201,
D-4600 Dortmund, West Germany

I. INTRODUCTION

The neuromuscular junction is the prototype of a rapid communication system between living cells. It has long been a preferred study object because 1) it is located at the body's periphery, 2) it is available in sufficient quantities, and 3) it can be studied with a large variety of techniques at the cellular, the subcellular, and the molecular levels. No other intercellular communication system has been studied in comparable detail and the neuromuscular junction thus probably offers the best possibilities to unraveling the basic principles of chemical excitation.

As usual, research on the excitation of muscle cells has progressed from purely phenomenological to detailed structural and functional studies of its major components. Because of the subsecond time range of the initial events, the time-resolved studies of muscle excitation have long remained an exclusive domain of electrophysiology while biochemistry has largely contributed structural information on the components involved. More recently, however, time-resolved biochemical studies have developed to the extent that they can provide independent bodies of evidence for the molecular mechanism of cholinergic excitation. It thus appears now possible to use the electrophysiological and biochemical findings in combination to develop minimal models of cholinergic excitation.

In this chapter we will first summarize the most prominent electrophysiological findings and define their biochemical correlates. We will then shortly review 1) direct measurements of cholinergically induced ion fluxes, 2) ligand binding studies with membrane-bound and purified receptor, and 3) kinetic studies involving cholinergic agonists and the acetylcholine receptor. We will limit ourselves to considering studies bearing on the mechanism of cholinergic excitation, and will concentrate on representative examples rather than reviewing each contribution in detail. By proceding from the cellular to the molecular level, the limitations but also the strengths of the biochemical approach will become evident. The final result is a detailed reaction scheme of cholinergic excitation that identifies particular agonist-receptor complexes as responsible for particular electrophysiological phenomena.

II. THE ELECTROPHYSIOLOGY OF CHOLINERGIC EXCITATION

Pharmacological studies on the neuromuscular junction were instrumental for the formulation of the receptor hypothesis [1, 2]. The concepts of chemical excitation [3, 4] and of ion channels [5, 6] were based to a large extent on electrophysiological studies involving muscle fibers. The initial processes of transmitter binding and specific conductivity change of the postsynaptic membrane appear to be of the same kind for vertebrate muscle and electroplaques.

The latter is degenerated muscle tissue in the sense that it is not capable of spontaneous contractions. Electrophysiological studies have provided the following major findings.

A. Characteristics of Changes in Membrane Potential

The changes in membrane potential induced by acetylcholine and its agonists are due to the opening of transmembrane channels located in the synaptic area of the postsynaptic cell. These channels transiently increase the membrane conductivity of sodium and potassium ions. The conductivity change can be explained on the basis of a single type of channel having an effective pore size of 6.4 Å [7]—ie, conducting equally well all ions of smaller diameter.

Autocorrelation analysis of the signal noise [for review see 8] provided the first evidence on individual channels. According to these studies [9–12], the closing rate of the agonist-activated channels is of the order of milliseconds but depends on the ligand applied. The channel opening rates increase with increasing agonist concentration whereas the closing kinetics are independent of the agonist concentration applied. Approximately 10^7 ions were calculated to pass through each channel in the course of an excitatory event, resulting in a mean conductance of 20 pS. Each channel contributes by about 0.3 μV to the depolarization of the membrane.

By reducing the area of observation [6], the direct measurement of single channel events became feasible. These studies reinforced and expanded the earlier studies of membrane noise: The rates of channel opening strongly depended on the concentration of agonist applied; the closing rates showed only a minor concentration dependence [6, 13–15]. Extrasynaptic channels had severalfold longer mean open times than synaptic ones (frog).

These findings are consistent with the concept of two components, a recognition site (receptor) and an effector site (channel) each residing in one or together in one membrane component. It has been recently established that both receptor and ion channel are integral parts of the protein moiety of the receptor [16].

B. Desensitization

Longer periods of exposure of muscle cells to acetylcholine lead to a decrease with time of the conductivity [17–21]. This effect is not due to breakdown of the transmitter because the conductivity cannot be increased by repeated application of acetylcholine. The effect can be reversed, however, by washing the cells. This phenomenon is called desensitization; it differs in intensity and time course for different agonists, and it is observed with most preparations including electroplaques from Electrophorus electricus.

From studies of single channel events [15], at least two types of desensitization differing in their rate constants are indicated.

These findings indicate the existence of two alternative pathways for the inactivation (closing) of the receptor-controlled channel: Inactivation may lead to a form that is rapidly reactivated in the presence of acetylcholine or to another one that cannot be reactivated immediately. Because depolarizing agents are required for the desensitized form, it is suggestive to assume that 1) the receptor is involved, and 2) the transmitter or its agonist is bound more tightly to the desensitized receptor than to the active form. This, then, suggests a reaction sequence from free, through complexed and active (ion-translocating) to complexed and inactive (desensitized) receptor.

C. Pharmacological Classification of Cholinergic Ligands

The ligands of the acetylcholine receptor can be subdivided into three groups: 1) Agonists induce similar responses as acetylcholine. 2) Antagonists compete with agonists for receptor binding but do not evoke the responses induced by agonists. The curaremimetic toxins [22] form a subclass of antagonists in that they are polypeptides and bind with exceptionally high affinity and dissociate only very slowly from the receptor. These properties have made them valuable tools for the detection and purification of acetylcholine receptors [23–27]. 3) Local anesthetics modify the properties of the agonist-activated channel without directly competing with the binding of cholinergic ligands to the receptor [28–33]. Cationic local anesthetics have been studied in more detail. According to the current view [8], they "transiently enter, plug and prevent the closure of the open endplate channel." Unfortunately, the above classification is rather idealized: Many ligands of the receptor have "mixed" properties—ie, they act as antagonists at low and as weak agonists at high concentrations (depolarizing antagonist). Others appear to be agonists at low concentrations but to increase the channel life time at higher concentrations (direct channel blockers).

The existence of cholinergic ligands and local anesthetics defines two functional sites, one of which (the local anesthetic site) probably resides at the channel. The overlapping activities (eg, decamethonium) indicate either or both of the following: similarities in the structural requirements for the binding to both sites, and close proximity of the two sites.

D. Agonist Efficacies

According to the classical occupation theory, drugs which act on one type of receptor must either be agonists or antagonists. Within this frame, all agonists should therefore produce the same maximal response. In practise, however, some drugs produce smaller maximal responses than others. Such compounds are called "partial agonists"; they are assumed to possess a smaller "intrinsic activity" [34, 35] or "efficacy" [36] than those with full maximal response.

Ariens' concept [34] leaves the basic proportionality between the fraction of receptors occupied and the response untouched. In contrast, in Stephenson's concept [36] different agonists are assumed to occupy different fractions of receptor to produce their maximal response. More direct studies of receptor-ligand interactions may clarify whether one of these concepts applies to the nicotinic receptor. The following explanations appear to be noteworthy alternatives: 1) There may exist several forms in equilibrium of saturated receptor with only one of them representing the active (ion-translocating) state. Different agonists may affect this equilibrium differently; 2) There may exist additional reaction steps between receptor saturation and response.

E. Sigmoid Dose-Response Curves

If the response is plotted versus the agonist concentration (and not the logarithm as before), again sigmoid curves are obtained [18, 37, 38]. The sigmoid shape of these curves is generally accepted, but there exists quantitative variation due to desensitization. This is particularly evident at high agonist concentrations [17–20]. An empirical measure for the degree of sigmoidity of these plots is the Hill coefficient n_H. It is obtained as the slope of a plot of log (response/(maximal response − response)) versus the logarithm of the agonist concentration. The resulting plot is of slope 1 if binding of a single agonist molecule per receptor causes the response [39]. The Hill coefficient is two if two agonists have to bind simultaneously to the receptor $(R + 2A \rightleftharpoons A_2R)$ to induce the response. For the more realistic case of a sequential saturation of the receptor with only the saturated receptor inducing the response $(R \rightleftharpoons AR \rightleftharpoons A_2R)$, the Hill plot would begin with a slope of two approaching a slope of 1 at very high agonist concentrations [40]. Noninteger values for the Hill coefficient are consistent with the sequential saturation model [40] and also with allosteric models [41–43].

Sigmoid dose-response curves may be caused by the following types of binding curves: 1) Given the appropriate proportionality function, any binding curve can be transformed into a sigmoid dose-response curve. 2) Assuming response to be linearly proportional to binding, binding to two sites (or the involvement of two distinct protomers) suffices to produce a sigmoid dose-response curve. On the basis of occupational models, response must then be assumed to be proportional to the saturation of two or more receptor sites; on the basis of allosteric models, two receptor sites are also required [41–43].

F. Chemical Modification and Response

Treatment of electroplaques [44, 45] or frog muscle cells [46] with the disulfide-reducing agent dithiothreitol (DTT) results in a reduced response of the preparation to acetylcholine and carbamylcholine. However, no change in the maximal response (efficacy) is observed. Furthermore, the antagonist hexamethonium acts as agonist at the DDT-treated preparation.

Because the effects observed in the presence of DTT are simple and clear-cut, the modification of a single functional site can be assumed. The modification by disulfide-reducing agents may therefore provide an additional criterium for the structural integrity of purified preparations.

G. Reconstitution of a Functional End Plate

Recently [16], purified receptor protein from Torpedo electric tissue has been incorporated into bilayers of a synthetic phospholipid. This system functioned similarly to a muscle endplate even below the phase transition temperature of the lipid employed: 1) The single channel events observed had similar mean open times and amplitudes as a reconstituted system consisting of Torpedo membrane patches grafted into the lipid bilayer and also as a mammalian muscle preparation. 2) The typical events of desensitization and the appearance of resensitization bursts were observed. 3) The single channel events showed the expected pharmacological specificity, the usual modulation of channel properties by a local anesthetic, and the established current-voltage relationship.

These finding establish that the receptor-controlled channel is part of the receptor itself and that the surrounding lipid serves largely if not exclusively as an ion-permeation barrier and a matrix for the receptor. Thus, the bio-electric properties of the postsynaptic membrane are determined by the properties of their proteins. On a different level, these findings provide strong support for the application of purified receptor preparations in biochemical studies.

III. ION CONDUCTIVITY OF END-PLATE MEMBRANES

Electrophysiological studies deal with changes of electrochemical potentials, and consequently they only permit inferences on the underlying ion fluxes. In principle, these fluxes can be measured directly by employing vesicular membrane preparations ("microsacs") or reconstituted systems consisting of planar lipid bilayers and the membrane components required for ion translocation: Microsacs are formed from receptor-rich membrane fragments in the presence of Ca^{2+} [47]. They can be incubated ("loaded") with radioactive alkali ions, rapidly washed, and dispersed in a nonradioactive medium. The rate of ion equilibration between internal and external medium (efflux) can then be measured. This rate is increased in the presence of cholinergic agonists [47, 48]. If the increase in rate is plotted versus the concentration of agonist applied, a sigmoid relationship is found [48]. As a consequence of the low time resolution of the earlier measurements, the data obtained in this way relate to the desensitized receptor. By employing rapid quench techniques [49–52], the rate of desensitization (slowing down of the efflux) is now experimentally accessible [53, 54].

As a reconstituted system, partially purified receptor has recently been incorporated into vesicles of artificial lipids [55]. There preparations were employed in efflux studies and yielded similar results to the natural microsacs. Planar lipid bilayers provide an even better system to study the ion conductivities of excitable membranes [56]. They can serve as supports for natural membrane patches or as lipid matrices into which other membrane components are dispersed. Using systems of this kind, the conductivity changes of the postsynaptic membrane have been shown to be caused by the acetylcholine receptor protein [16]. The other membrane components (lipids, proteolipids) appear to have only small if any modulating effects on the electric properties of these membranes [16].

Taken together, the ion flux measurements provide the molecular basis for many electrophysiological phenomena. In spite of the limitations of the flux methods (limited reproducibility, insufficient characterization of the components, basal ion fluxes, and membrane noise), there is little doubt that 1) the transient changes of membrane conductivity relate to true ion fluxes along preestablished concentration gradients through membrane pores, and 2) the receptor protomer (a protein of apparent molecular weight 300,000 [57] and composed of five subunits ($\alpha_2\beta\gamma\delta$)) is the basic receiving and transmitting unit of the end plate.

IV. BINDING STUDIES

So far we have discussed only the response of nicotinic end plates. We shall now consider the reaction immediately preceding the response.

The interaction of the acetylcholine receptor with its ligands has been studied by a variety of methods employing whole cells, membrane preparations, and solubilized receptor. The studies with whole cells are limited by the low molar concentrations of receptor. They therefore require radioactive ligands of very high specific activity and high affinity of binding. Employing iodinated α-bungarotoxin and cultured chick embryo myogenic cells, the concentrations of receptor [58] and some basic properties of the ligand-binding sites [58, 59] were studied. Largely because of the lack of sufficient material, these studies are too cumbersome to permit a detailed analysis of the mode of binding of toxin and other ligands to the receptor.

Receptor-containing membrane preparations and purified receptor from electric tissue can overcome the problem of limited concentration. Because the natural membrane environment of the receptor is preserved in the membrane preparations, they are less artificial than the detergent-solubilized receptor. As a consequence, experimental findings obtained with the solubilized receptor always carry some ambiguity as long as they are not supported by additional evidence from other sources [60]. At the same time, however, the solubilized receptor provides a preparation that can be handled with the same ease as a water-soluble enzyme preparation and that permits straightforward interpretation of the experimental data [61].

The equilibrium binding studies with whole cells were exclusively performed with solid-phase assays [58, 62, 63]; those with membrane fragments were performed with equilibrium dialysis [64–67], centrifugation [68–73], filtration [69, 74–76], or spectroscopic methods [77–81]. Except for centrifugation, similar methods can be applied to the studies with the solubilized receptor. The advantages and disadvantages of these methods have been discussed in detail elsewhere [60]. In short, equilibrium dialysis is theoretically the most rigorous method, but is restricted by nonspecific binding to the dialysis membranes and by the excessively long periods of time required to achieve transmembrane equilibrium. In these studies the viability and structural integrity of the proteins employed always remains in question. The centrifugation assay is considerably faster than and just as straightforward as equilibrium dialysis. Because a large variety of tube material is available, nonspecific absorption can be minimized [68]. This assay is therefore the method of choice for binding studies with membrane preparations.

The filtration assays permit a rapid separation of bound from free ligand. At the same time, however, the binding equilibrium is perturbed by this separation. As a rule, these assays should be used only for the determination of complexes with dissociation rates significantly smaller than the filtration rate. Filtration assays are therefore limited to studies involving the slowly dissociating polypeptide neurotoxins. In studies with small cholinergic ligands [69] they are necessarily of only little significance.

Spectroscopic methods, finally, are limited to reaction components with the required spectral properties. The intrinsic fluorescence of the receptor protein although the most direct probe, unfortunately, is only very weak [82, 83]. In addition, it is changed only by a few percentage points in the presence of cholinergic ligands [78, 83]. Because the natural transmitter is nonfluorescent, spectroscopic methods require then either chemically modified receptor or ligands of the wanted properties. As has recently been discussed [84], only a few of these probes introduce sufficiently small perturbations to permit studies of receptor interaction with the wanted significance.

There are two serious problems that complicate binding studies with the cholinergic system and their interpretation: 1) Most receptor preparations contain significant amounts of other proteins (eg, acetylcholine esterase) that display similar affinity for cholinergic ligands as the receptor itself. This problem is amplified by the fact that the concentration of esterase matches or exceeds that of the receptor in unwashed membrane homogenates or cellular preparations. Repetitive salt washes reduce the content of contaminating proteins, but only a few purified preparations of receptor can be considered essentially free of them. The problem arises from the fact that all cholinergic ligands, with the notable exception of the polypeptide toxins, bind to both receptor and esterase. Consequently, only toxin affinity columns (and not those with other ligands) lead to an enrichment of receptor with respect to the esterase content. Similarly, only with toxin can the total concentration of receptor be determined in the presence of acetylcholine esterase. These properties make the polypeptide toxins a necessary requirement for both receptor assays and receptor purification [85]. 2) The most serious limitation of binding studies with the cholinergic system concerns their significance with respect to physiological observations. The active state of the end plate (and the receptor) is only transient and short-lived. Equilibrium binding data therefore contain large contributions from inactive states.

A. Binding to the Membrane-Bound Receptor

Several reports have appeared over the years on the interaction of the natural transmitter with membrane fragments from electric tissue. The most striking difference in the results of these and electrophysiological dose-response curves is the range of concentrations required to reach half-maximal saturation and half-maximal response, respectively. These concentration ranges differ by at least two orders of magnitude, the range for half-maximal binding being shifted to the lower concentrations. The effect has been attributed to desensitization; ie, the desensitized receptor is assumed to bind acetylcholine (and its agonists) with several orders of magnitude higher affinity than the active receptor. In view of the ease by which sigmoid dose-response curves are observed, it is further remarkable that sigmoid binding curves are observed only under very limited conditions [67–74]. Because both the concentration range and shape of the binding curves are only reminiscent of rather than identical to the related dose-response curves, the correlation between the two is apparently not simple. The most recent and most detailed study of acetylcholine binding [68] has determined the boundary conditions for sigmoid acetylcholine binding curves to the membrane bound receptor: 1) The membrane preparations must be obtained in the presence of appropriate chelating, sulfhydryl blocking, and active serine blocking agents. 2) The membrane preparations must be relatively fresh (only a few days old). 3) The assay must be performed under conditions permitting an unusually low range of error.

Under these conditions, weak positive cooperativity of two classes of acetylcholine-binding sites at the receptor is observed. The small differences in K_D values of the two classes of sites ($K_{D1} = 25$ nM; $K_{D2} = 8$ nM) provide the rational for the limited reproducibility and the contrasting findings of earlier studies. In conclusion, positively cooperative binding of two acetylcholine molecules per receptor molecule appears likely. Since cooperative binding is a special case of a sequential mode of interaction, this ordered mode of association and dissociation is the most central finding of these studies.

The binding of agonists of acetylcholine has been studied in much lesser detail. This is mainly caused by the fact that these ligands often exhibit additional pharmacological effects. These effects limit the significance of such studies. In contrast, the results of binding studies with cholinergic antagonists are in full agreement with electrophysiological studies [for review see 86]: There is no indication of sigmoidicity in the binding curves [69], and the concentration for half-maximal binding agrees with the electrophysiologically determined concentration range of the blocking effect caused by these ligands.

B. Binding to the Purified Receptor

Binding studies to the purified receptor are based on the following considerations: 1) Because of the limited reproducibility and the resulting ambiguity in results of studies with membrane preparations (see above), a simpler, more defined preparation was required. 2) The solubilized receptor permitted the study of properties of the protein in the absence of any membrane components. However, the purified receptor also required modifications of the commonly applied assays: The most advantageous assay—ie, centrifugation—requires excessively large g values, and hence has not been applied. Because of adsorption to the membranes and inactivation, equilibrium dialysis is as problematic as with the membrane preparations. Similarly, filter assays again are too slow compared to the rates of ligand dissociation to study the binding of small ligands directly. Consequently, the most significant studies with the purified receptor [85] were performed with radioactively labeled α-neurotoxins and the filter assay. Employing this assay procedure, the binding properties of small cholinergic ligands were then deduced from competition binding curves with the neurotoxins [85]. The following results were obtained: 1) There was complete agreement in the concentration range of antagonist binding to the purified receptor and of blocking at the cellular level. 2) Small cholinergic agonists and antagonists differed in their mode of binding: Antagonists bound to a single class of sites, and agonists bound to two classes of sites (but the same number of sites as the antagonists) at the receptor. The binding pattern of agonists suggested negatively cooperative interactions of agonist sites. 3) Agonist binding again differed in the concentration range from the electrophysiological dose-response curves.

These findings, although obtained indirectly, pointed strongly to a similar if not identical mode of recognition at the cellular (membrane-bound receptor) and the molecular (solubilized receptor) level. Employing a new, rapid, and true equilibrium assay, the preservation in the solubilized state of the recognition function has recently been demonstrated [87]. In these studies, a fluorescent agonist of acetylcholine (NBD-5-acylcholine*) with closely similar pharmacology was employed. With the purified receptor from Electrophorus electricus, binding of this ligand was observed as full quenching of the ligand's fluorescence. The large size and simplicity of the fluorescence effect made the analysis of binding data particularly straightforward. In addition to direct binding studies, competition studies with representative nonfluorescent agonists and antagonists were performed. The following results were obtained [87]: 1) All ligands studied competed for the same number of sites; 2) antagonists bound to these sites with only one affinity; 3) agonists (including the monitoring ligand NBD-5-acylcholine) bound to half of the sites with high affinity, and to the other half with markedly lower affinity; and 4) in the presence of disulfide-reducing agents, the binding patterns of some agonists and antagonists were changed in accordance with the electrophysiological changes observed under the same conditions with electroplaques [44, 45].

Taken together, the binding studies show a remarkable quantitative stability of the receptor's recognition function from the cellular to the molecular level of purification. Besides stressing the protein moiety as the site of function, the adequacy of the solubilized, affinity-purified receptor as preparation for quantitative functional studies is indicated. Since the most obvious function of the receptor (initiation of the primary response) relates to short-lived states, equilibrium binding studies necessarily have only limited significance in their analysis. A full understanding of the mechanism of ligand recognition therefore requires a kinetic analysis. By the equilibrium studies, the purified receptor has been indicated to be a suitable preparation for such purposes.

V. KINETIC STUDIES

The kinetics of ligand interaction with the acetylcholine receptor have recently been reviewed in detail [84]. The only new study not covered in this review concerns the interaction of NBD-5-acylcholine and the purified receptor [88, 89]. This study produced a set of data sufficient to develop and test reaction schemes of cholinergic excitation. In contrast, earlier studies of this kind [90–95] employed preconceived models for the analysis of data. Here we shall first discuss the new study [88] and then relate it and other kinetic data to the established findings of muscle electrophysiology.

---

*NBD-5-acylcholine: N-7-(4-nitrobenzo-2-oxa-1,3-diazole)-5-aminopentanoic acid β(N-trimethylammonium)ethyl ester.

A. Kinetics With the Fluorescent Agonist NBD-5-Acylcholine

The interaction of NBD-5-acylcholine and the affinity-purified receptor was studied by means of a laser stopped-flow system. Receptor from electric eel was chosen because quenching of NBD-5-acylcholine in the bound state is complete [87]. A stopped-flow system was used because 1) it permits to observe the full course of the reaction, and 2) it permits the application of nonliganded ("virgin") receptor. In contrast, the frequently employed relaxation techniques [92] always originate from partially or fully occupied receptor and, thus, with the cholinergic system relate to desensitized forms of the receptor. The complete quenching of the drug's fluorescence in the bound state permitted the calibration of the fluorescence signal in terms of free ligand concentration. With this simple relationship an unambiguous analysis of data was achieved.

The association reaction was studied after rapid mixing of free receptor and fluorescent ligand. The association kinetics consisted of several steps extending over several orders of time. From the initial rates of association, a lower limit for the second order forward rate constant of 2×10^8 M^{-1} s^{-1} was calculated for physiological salt conditions and 20°C. This classifies the initial association reaction as being close to diffusion-controlled [96].

The results of the dissociation kinetics of preformed NBD-5-acylcholine–receptor complexes provided direct insights into the reaction mechanism: 1) When dissociation was initiated by dilution, biphasic kinetics were observed under all conditions. This even applied to reaction conditions with exclusively monoliganded receptor present (ie, low receptor saturation). Thus, the dissociation of monoliganded receptor involves *two* reaction steps. In view of the equilibrium-binding studies with the same system [87], this then requires the assumption of two distinct populations of complexes contributing to the first class of agonist binding sites. 2) When dissociation was initiated by addition of a large excess of nonfluorescent ligand, a very different reaction pattern was observed: Dissociation was altogether much slower with the half-life of fluorescent complexes increasing with increasing concentration of competing ligand. This result requires the assumption of a strictly ordered ("first on–last off") mode of association to and dissociation from the two ligand-binding sites at the receptor. In particular, it does not permit any isomerization of sites as was observed for toxin dissociation [85]. Thus, if the fluorescent ligand occupies only the first receptor site and the second site is then filled with the nonfluorescent ligand, dissociation of the fluorescent ligand is necessarily slowed down. Because of the strict order of dissociation ("last on–first off"), the fluorescent ligand can only dissociate if the second site of the receptor is again unoccupied (the nonfluorescent ligand has to dissociate first).

The basic phenomena observed in the kinetic studies with NBD-5-acylcholine therefore already expanded the picture obtained from the equilibrium binding studies: 1) Association to and dissociation from the two sites previously established [87] is strictly ordered. 2) There exist two forms (two types of receptor-ligand complexes) in equilibrium of the monoliganded receptor.

At the second level of analysis, the kinetic data were numerically fitted to the models that can be imagined on the basis of the more qualitative findings. Considering two types of monoliganded receptor and a strictly ordered mechanism of formation and breakdown of the monoliganded and diliganded receptor, the following five minimal schemes were proposed [88]:

$$
\begin{array}{cccc}
 & AR & \overset{3}{\rightleftharpoons} & AR\text{-}A \\[2pt]
R \quad \underset{2}{\overset{1}{\rightleftarrows}} & & & \\[2pt]
 & RA & \overset{4}{\rightleftharpoons} & A\text{-}RA
\end{array}
\qquad (1a)
$$

$$
\begin{array}{cccc}
 & AR & \overset{3}{\rightleftharpoons} & AR\text{-}A \\[2pt]
R \quad \underset{2}{\overset{1}{\rightleftarrows}} & \updownarrow 5 & & \\[2pt]
 & RA & \overset{4}{\rightleftharpoons} & A\text{-}RA
\end{array}
\qquad (1b)
$$

$$
\begin{array}{cccc}
R \overset{1}{\rightleftharpoons} & AR & \overset{3}{\rightleftharpoons} & AR\text{-}A \\[2pt]
 & \updownarrow 5 & & \\[2pt]
 & AR^{\times} & \overset{4}{\rightleftharpoons} & AR^{\times}A
\end{array}
\qquad (2a)
$$

$$
\begin{array}{cccc}
R \overset{1}{\rightleftharpoons} & AR & \overset{3}{\rightleftharpoons} & AR\text{-}A \\[2pt]
 & \updownarrow 5 & & \updownarrow 6 \\[2pt]
 & AR^{\times} & & AR^{\times}A
\end{array}
\qquad (2b)
$$

$$
\begin{array}{cccc}
R \overset{1}{\rightleftharpoons} & AR & \overset{3}{\rightleftharpoons} & AR\text{-}A \\[2pt]
 & \updownarrow 5 & & \updownarrow 6 \\[2pt]
 & AR^{\times} & \overset{4}{\rightleftharpoons} & AR^{\times}A
\end{array}
\qquad (2c)
$$

All five schemes consider the same number of molecular species; they differ only in the reaction pathways by which these molecular species are connected with each other: Schemes 1 consider two structurally different sites at the receptor that are filled in parallel. (Note that schemes 1 require also the assumption of *two* forms of the diliganded receptor in order to allow for a strictly sequential mechanism of interactions). In schemes 2, only one type of monoliganded receptor is initially formed; this complex can undergo a conformational change to a second form. (Note that schemes 2 can be "doubled" to account for the formation of *two* forms of monoliganded receptor from free receptor. Each of these would then undergo a conformational change to another form of monoliganded receptor, and the two forms each would combine with more ligand to form a total of four different forms of diliganded receptor. Such expanded schemes would not affect the fits but the resulting rate constants would be half those of the original schemes.)

It is important to realize that the five schemes consider *four* ligand-receptor *complexes* but only *two sites*. The existence of two sites at the receptor was shown by both equilibrium binding [87]* and kinetic [88] studies. That more than two complexes with these two sites are involved was the result of the kinetic studies alone [88].

By fitting several sets of association and dissociation kinetics simultaneously [88], numerical values were attached to the various reaction steps proposed in the schemes. Independent of the particular reaction scheme applied, these exposed additional key features of the molecular mechanism of cholinergic excitation: One of the saturated receptor-ligand complexes (AR-A) is rapidly formed and dissociated, the other (A-RA or AR*A) forms more slowly and dominates at equilibrium. These properties invite comparison with the results of electrophysiological studies of muscle end-plates: The agonist-induced ion channel is rapidly activated [6, 9–15] and requires the binding of two agonist molecules per receptor [40, 98]. This state of the receptor is short-lived [6, 9–15]. The concentration of active states decreases with the time of exposure to agonists ("desensitization") [15, 16]. With the assumption of AR-A being the active form and AR*A (or A-RA) being the desensitized form of the receptor, basic agreement between biochemical and electrophysiological data is achieved. This is particularly striking when the data for acetylcholine are considered (Table I). These were obtained by analyzing the dissociation kinetics of NBD-5-acylcholine–receptor complexes observed in the presence of various concentrations of acetylcholine [88]. The key findings were: 1) The complexes AR and AR-A are formed very rapidly (k_1, k_3). 2) The complex assumed to be the active state (AR-

---

TABLE I. Resulting Reaction Parameters Derived for Acetylcholine From Chase Experiments With NBD-5-Acylcholine-Receptor Complexes [88]

Parameter	Reaction scheme	
	2a	1a
$k_1 (M^{-1} s^{-1})$	5×10^8	1×10^8
$k_{-1} (s^{-1})$	740	150
$K_{D1} (M)$	1.5×10^{-6}	1.5×10^{-6}
$k_2 (M^{-1} s^{-1})$		4.3×10^7
$k_{-2} (s^{-1})$		8.3
$K_{D2} (M)$		2×10^{-7}
$k_3 (M^{-1} s^{-1})$	5×10^6	1×10^8
$k_{-3} (s^{-1})$	400	1,340
$K_{D3} (M)$	8×10^{-5}	1.3×10^{-5}
$k_4 (M^{-1} s^{-1})$	1.5×10^{-5}	5.8×10^4
$k_{-4} (s^{-1})$	0.09	0.1
$K_{D4} (M)$	6×10^{-7}	1.8×10^{-6}
$k_5 (s^{-1})$	170	
$k_{-5} (s^{-1})$	20	
K_{D5}	0.12	
$K_D(1)^*$	1.6×10^{-7}	1.8×10^{-7}
$K_D(2)^*$	6.7×10^{-7}	2.0×10^{-6}

*$K_D(1)$ and $K_D(2)$ are the K_D-values of the two classes of binding sites observed at equilibrium [87]. They relate to the K_D-values of the individual reaction steps as described in ref. [88].

A) is short-lived with decay constants (k_{-3}) of the same order as the mean open time of the acetylcholine activated channel [6, 13–15]. 3) The K_D value of AR-A is of the same order as the concentration required for half-maximal response of the electroplaque [12, 38, 60, 86]. 4) High levels of AR-A exist only transiently because AR*A (or A-RA) is favored at equilibrium.

B. Comparison With Other Kinetic Studies

Because the binding and kinetic studies with NBD-5-acylcholine [87, 88] are the most detailed yet reported, comparison with other studies is limited to a few reaction parameters: The association kinetics of dansyl-6-acylcholine and membrane fragments from Torpedo marmorata [90, 91] also consisted of several reaction steps. The initial rapid fluorescence change yielded a second-order rate constant of $9.5 \times 10^7 M^{-1} s^{-1}$, which closely agrees with the data for NBD-5-acylcholine and purified eel receptor. Heidmann and Changeux [90, 91] also observed a second bimolecular reaction step. This step, however, was considerably slower ($3.5 \times 10^5 M^{-1} s^{-1}$) than the one observed in our study ($> 10^8 M^{-1} s^{-1}$), and therefore most likely relates to a reaction step that does not exist with the eel receptor or only is observed with membrane fragments. Other studies [83, 92] report only a single bimolecular association step of the order of 10^7 M^{-1} s^{-1}. In the first instance [83], these findings are based on the very small changes of the intrinsic fluorescence of whole membrane fragments. In the latter instance [92], the interaction of Ca^{2+} with murexide following displacement from the receptor by acetylcholine was monitored. The rate constant observed in both studies therefore most likely relates to conformational changes of the receptor following the initial association reaction [84].

Slow processes of first order are observed in the association kinetics with both NBD-5-acylcholine [88] and dansyl-6-acylcholine [90, 91]. They probably relate to the slow processes observed in both electrophysiological and biochemical studies of receptor-ligand interaction [15, 18]. Studies by Quast et al [93] and Dunn et al [94] indicate several consecutive processes in the time range of seconds after the initial complex between agonist and membrane-bound receptor has been formed.

The studies with NBD-5-acylcholine [88] and with dansyl-6-acylcholine [90, 91] both indicate rather complicated dissociation kinetics of agonist-receptor complexes. Much simpler, but also more limited in their scope, are the results obtained by Barrantes [95] and Neumann and Chang [92] on the dissociation of suberyldicholine from membrane-bound Torpedo receptor [83] and of acetylcholine at pH 8.5 from purified torpedo receptor. These studies identify single first-order rate constants of 10 s^{-1} [95] and 140 s^{-1} [92], respectively. They are probably composed of several of the dissociation rate constants listed in Table I.

The considerable discrepancies in the results of the kinetic studies with the cholinergic system may be attributed to 1) the different preparations applied (membrane fragments from Torpedo [90, 91, 95], purified receptor from Torpedo [92], or Electrophorus [87, 88]); 2) the different experimental methods (intrinsic fluorescence of membrane fragments [83], extinction of murexide [92], fluorescence of acetylcholine analogues [87–91, 95] and of probes with local anesthetic action [93] or of indirect probes [94]); and 3) the different procedures of analysis of the kinetic data. The latter should not be underestimated in its influence on the results obtained: The quantitative values for rate and equilibrium binding constants strongly depend on the reaction scheme (model) applied. A proper comparison of the reported results would therefore require their recalculation with the same model assumptions. Furthermore, most of the reported kinetic data were analyzed by assuming a sum of exponential decays:

$$f(t) = \sum_{i=1}^{n} A_i \times e^{-k_i t} \tag{3}$$

This kind of analysis is correct only if each reaction step is independent of the others, or (as in the case of relaxation kinetics) if only minor disturbances of the equilibrium are considered. Since association to and dissociation from the receptor are strictly ordered (sequential) processes [88], the above procedure of data analysis (equation 3) must lead to erroneous results. Thus, it is probably due mainly to the differences in the analysis of data that similar experimental findings have been reported as supporting rather different models. Conversely, results that appear inconsistent with each other have been cited as supporting the same models.

VI. ON THE MECHANISM OF CHOLINERGIC EXCITATION

Both the recent physiological and the biochemical studies of muscle excitation have the ultimate goal of elucidating the molecular mechanism by which extracellular chemical signals are translated into electrical and intracellular mechanical events. This final chapter is therefore aimed at assessing the related findings from both disciplines.

A. The Stoichiometry of Sites

Electrophysiological studies have long indicated the requirement of at least two molecules of acetylcholine bound per receptor to gate the receptor-controlled ion channel [40]. Recent equilibrium-binding studies with membrane-bound [68–70, 74, 99] and solubilized [87] receptors again pointed to two binding sites per receptor.

B. The Shape of Dose-Response Curves and Binding Curves

Dose-response curves of the cholinergic system generally are sigmoid in shape [12, 38, 40, 98]. In contrast, sigmoid binding curves of acetylcholine and the receptor are observed only at very limited experimental conditions [68, 69, 73]. This discrepancy does not appear very critical for the following reasons: 1) Because of the excessively long period of incubation with agonists, all equilibrium binding studies relate to desensitized receptor (see below). It is likely that the properties of the desensitized receptor differ from those of the active receptor. 2) Sigmoid dose-response curves can also be explained on the basis of hyperbolic binding curves [87, 97].

C. The Mode of Binding

Both the voltage-jump relaxation studies on isolated electroplax [12, 100] and the laser stopped-flow studies with purified receptor [88] indicate a strictly ordered ("first on–last off") mechanism of agonist association to and dissociation from the receptor. A sequential mode of binding is also indicated from the positively cooperative binding of acetylcholine to the membrane-bound receptor [68].

D. The Affinities of Binding

Independent of whether membrane fragments or purified receptor was employed, the binding affinities always differed by orders of magnitude from the concentrations of agonist required to elicit half-maximal response. Thus, the agonist-receptor complexes dominating the equilibrium binding curves most likely differ from those that cause the response. Consequently, two types of fully saturated agonist-receptor complexes can be assumed—one of them active but existing only transiently, the other inactive and dominating at equilibrium.

E. The Kinetics of Activation and Desensitization

Both the kinetics of agonist association to and dissociation from the receptor indicate several reaction steps [88–91]. In addition, the dissociation kinetics require the assumption of two forms of the monoliganded receptor [88]. As a consequence, the resulting reaction schemes (schemes 1, 2) are of sufficient complexity to accommodate the electrophysiological findings concerning receptor activation and desensitization. Desensitization is characterized by the following: It is a reversible process that requires sufficient periods of incubation of agonist and the receptor [15, 18, 19, 21]. It does not necessarily require receptor activation [21]. Breakdown of the desensitized form of saturated agonist-receptor complex does not necessarily proceed via the active form of receptor [21]. Based on these properties a branched reaction scheme involving two diliganded forms of receptor with independent pathways of formation and breakdown is required.

F. A Reaction Model Based on Kinetic Data

With the above considerations in mind, the following reaction scheme for the interaction of cholinergic agonists and the acetylcholine receptor can be drawn:

$$
\begin{array}{ccccc}
R & \underset{2}{\overset{1}{\rightleftharpoons}} & AR & \underset{}{\overset{3}{\rightleftharpoons}} & AR\text{-}A \\
 & & \updownarrow 5 & & \updownarrow 6 \\
 & & AR^* & \overset{4}{\rightleftharpoons} & AR^*A
\end{array}
$$

The scheme considers two binding sites at the receptor with two forms each of the monoliganded and diliganded receptor. It is proposed that one of the saturated complexes (AR-A) is the active (ion-gating) form, and that the other saturated complex (AR*A) is the inactive (desensitized) form of the receptor. The two forms of monoliganded receptor are both inactive. Thus, electrophysiological studies relate only to the formation and breakdown of AR-A whereas biochemical studies cover all reaction steps of the above scheme. In the following we analyze the above model in view of the available electrophysiological and biochemical data.

G. Rate Constants of Agonist Interaction

The rate constants resulting for the above model from the rapid kinetic studies with NBD-5-acylcholine and the receptor have been calculated [88]; they are listed in Table II. Although subject to a considerable range of error, these data may serve to illustrate the properties of the model. According to Table II, AR-A is rapidly formed (k_3) and dissociated (k_{-3}). Assuming AR-A to be the active (ion-gating) form of receptor, the association rate would then correlate with channel opening, the dissociation rate with channel closing. At the concentrations of acetylcholine assumed to exist in the synaptic cleft during an excitatory event [101], AR-A would indeed be formed within a few microseconds [12] with its half-life being of the order of a few milliseconds [6, 16, 101]. In contrast, AR*A would be formed more slowly but would increasingly dominate the equilibrium just as is established for the event of desensitization [18, 19, 21].

H. A Simulated Event of Cholinergic Excitation

The association of transmitter and receptor following synchronous release of transmitter into the synaptic cleft can be mimicked by rapid mixing experiments with fluorescent ligands [88]. In Figure 1 the concentration versus time plot for a simulated experiment of this kind for acetylcholine is shown. As can be seen, the kinetics are successively dominated by the complexes AR, AR-A, AR*, and AR*A. In terms of the proposed correlation, the concentration versus time function for AR-A would then mimic the conductivity changes of a voltage-clamped muscle cell [9] while the decrease in the concentration of AR-A with time is reminescent of the fast process of desensitization [15].

I. Is Channel Gating a Separate Event?

Although assumed in many reaction schemes [8, 11, 12, 18, 37, 49, 51, 84, 86, 98], we are not aware of any data that properly *require* the assumption of separate reaction steps for channel gating. In contrast, there is strong evidence that the rapidly formed diliganded form of the receptor (AR-A) *is* the conductive state: 1) The ion channel has been shown to be part of the protein moiety of the receptor [16]. 2) The decay constant of AR-A as determined in stopped-flow experiments [88] is of the same order as the mean open time of the agonist activated channel [6, 9–15]. 3) At the molecular level, attachment of a second agonist molecule (to form AR-A from AR)

TABLE II. Resulting Reaction Parameters of Simultaneous Fits of the Equilibrium Binding and Kinetic Data for Scheme 2c According to [88]

Parameter	Reaction scheme
$k_1(M^{-1}\ s^{-1})$	5.5×10^8
$k_{-1}(s^{-1})$	5.5
$K_{D1}(M)$	1.0×10^{-8}
$k_3(M^{-1}\ s^{-1})$	4.6×10^8
$k_{-3}(s^{-1})$	23.0
$K_{D3}(M)$	5.0×10^{-8}
$k_4(M^{-1}\ s^{-1})$	1.4×10^5
$k_{-4}(s^{-1})$	0.0084
$K_{D4}(M)$	6.0×10^{-8}
$k_5(s^{-1})$	2.0
$k_{-5}(s^{-1})$	0.7
K_{D5}	0.35
$k_6(s^{-1})$	0.83
$k_{-6}(s^{-1})$	0.35
K_{D6}	0.42
$K_D(1)(M)$	2.6×10^{-9}
$K_D(2)(M)$	5.7×10^{-8}

may just sufficiently change the charge distribution to "open" the receptor-integrated channel. Thus, in view of the close structural relationship between binding and response site [16] and the high speed of response [12], it appears logical to confine the formation of the conducting receptor state to only two reaction steps of second order.

VII. CONCLUDING REMARKS

The analysis of the electrophysiological and biochemical data on cholinergic excitation results in a plausible model (scheme 2c) that is surprisingly simple: 1) Receptor activation is a two-step mechanism with only the saturated (diliganded) receptor eliciting the response. 2) Receptor desensitization is a separate reaction branch linked to the monoliganded and diliganded receptor forms produced during activation. 3) By linking the response to only the diliganded receptor (AR-A), both the affinities and efficacies deduced from dose-response curves of agonists are easily understood: The agonist affinity

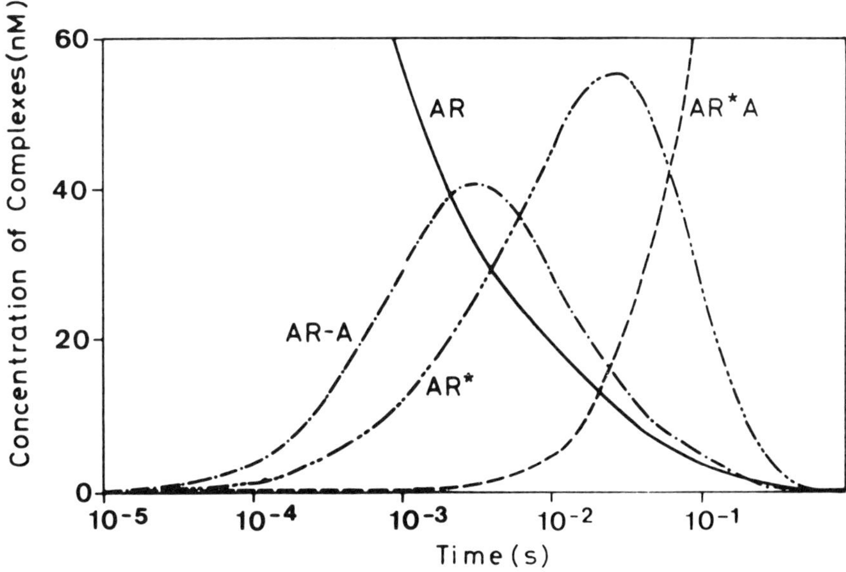

Fig. 1. Kinetics of the formation and breakdown of individual acetylcholine receptor complexes in the course of an association reaction. Initial concentrations: receptor, 10^{-7} M and acetylcholine 10^{-4} M. The data were calculated according to [88] on the basis of scheme 2a and Table I.

(concentration for half-maximal response) is determined by the K_D value of the complex AR-A, and the agonist efficacy may relate to the ratio of K_D values of AR-A and AR. 4) With the response elicited by AR-A, dose-response curves will be of sigmoid shape independent of the specific mode of binding to the receptor (and consequently, independent of whether the ligand binding curve is sigmoid or hyperbolic in shape). The related Hill plot will begin with a slope of 2 approaching 1 at high ligand concentrations. 5) The rates of channel opening and closing are determined by the rate constants of AR-A. Accordingly, channel opening is a rapid process depending on both the concentrations of monoliganded receptor and agonist whereas channel closing is a process of first order. 6) The mean open time of the channel is linked to the dissociation rate constant of AR-A. The total rate of dissociation of agonist from the receptor may be significantly slower [88, 89, 101]. 7) Desensitization can proceed from the monoliganded receptor and does not require the existence of the active complex AR-A. Furthermore, the inactive, saturated complex AR*A can decay to free receptor without the requirement of resensitization.

The model satisfies the basic findings from chemical kinetics and electrophysiology, and may thus serve as the basis for further probing into the molecular mechanism of muscle excitation.

ACKNOWLEDGMENTS

We are grateful for many stimulating discussions with our colleagues G. Fels, E. Wolff, L. Smart, and H. W. Meyers. Our work is supported by grants from the Deutsche Forschungsgemeinschaft, Schwerpunktprogramm "Molekulare Mechanismen zellulärer Signalaufnahme."

VIII. REFERENCES

1. Langley JN: On the reaction of cells and nerve-endings to certain poisons, chiefly as regards the reaction of striated muscle to nicotinic and to curari. J Physiol 33:374, 1905.
2. Clark AJ: "The Mode of Action of Drugs on Cells." Baltimore: Williams and Wilkins, 1933.
3. Hodgkin AL: The ionic basis of electrical activity in nerve and muscle. Biol Rev 26:339, 1951.
4. Eccles, JC, Katz B, Kuffler SW: Effect of exrine on neuromolecular transmission. J Neurophysiol 5:211, 1942.
5. Fatt P, Katz B: An analysis of the end-plate potential recorded with an intracellular electrode. J Physiol 115:320, 1951.
6. Neher E, Sakmann B: Single channel currents recorded from membrane of denervated frog muscle fibres. Nature 260:799, 1976.
7. Dwyer TM, Adams DJ, Hille B: The permeability of the end-plate channel in organic cations in frog muscle. J Gen Physiol 75:469, 1980.
8. Adams PR: Acetylcholine receptor kinetics. J Membrane Biol 58:161, 1981.
9. Magleby KL, Stevens CF: The effect of voltage on the time course of end-plate. J Physiol (Lond) 223:151, 1972.
10. Adams PR: Kinetics of agonist conductance changes during hyperpolarization at frog endplate. Br J Pharmacol 53:308, 1975.
11. Sheridan RE, Lester HA: Relaxation measurements on the acetylcholine receptor. Proc Natl Acad Sci USA 72:3496, 1975.
12. Sheridan RE, Lester HA: Rates and equilibrium at the acetylcholine receptor of Electrophorus electroplagues. J Gen Physiol 70:187, 1977.
13. Neher E, Sakmann B, Steinbach JH: The extracellular patch clamp: A method for resolving currents through individual open channels in biological membranes. Pflugers Arch 375:219, 1978.
14.. Sakmann B, Heesemann J: Recording of single channel currents from the subsynaptic membrane of frog neuromuscular junction. Pflugers Arch 382:34, 1979.
15. Sakmann B, Patlak J, Neher E: Single acetylcholine activated channels show burst-kinetics in presence of desensitizing concentrations of agonists. Nature 286:71, 1980.

16. Boheim G, Hanke W, Barrantes FJ, Eibl H, Sakmann B, Fels G, Maelicke A: Agonist-activated ionic channels in acetylcholine receptor reconstituted into planar lipid bilayers. Proc Natl Acad Sci USA 78:3586, 1981.
17. Fatt P: The electromotive action of acetylcholine at the motor end-plate. J Physiol (Lond) 111:408, 1950.
18. Katz B, Thesleff S: A study of the "desensitization" produced by acetylcholine at the motor end-plate. J Physiol (Lond) 138:63, 1957.
19. Larmie ET, Webb GD: Desensitization in the electroplax. J Gen Physiol 61:263, 1973.
20. Lester HA, Changeux J-P, Sheridan RE: Conductance increases produced by bath application of cholinergic agonists to Electrophorus electroplagues. J Gen Physiol 65:797, 1975.
21. Magleby KL, Pallotta BS: A study of desensitization of acetylcholine receptors using nerve-released transmitter in the frog. J Physiol 316:225, 1981.
22. Chang CC, Lee CY: Isolation of neurotoxins from the venom of Bungarus multicinctus and their modes of neuromuscular blocking action. Arch Intern Pharmacodyn 144:241, 1963.
23. Klett RP, Fulpius BW, Cooper D, Smith M, Reich E, Possani LD: The acetylcholine receptor. I. Purification and characterization of a macromolecule isolated from Electrophorus electricus. J Biol Chem 248:6841, 1973.
24. Schmidt J, Raftery MA: Purification of acetylcholine receptors from Torpedo californica electroplax by affinity chromatography. Biochemistry 12:852, 1973.
25. Meunier JC, Sealock R, Olsen R, Changeux J-P: Purification and properties of the cholinergic receptor from Electrophorus electricus electric tissue. Eur J Biochem 45:371, 1974.
26. Karlsson E, Heilbronn E, Widlund L: Isolation of the nicotinic acetylcholine receptor by biospecific chromatography on insolubilized Naja naja neurotoxin. FEBS Lett 28:107, 1972.
27. Eldefrawi ME, Eldefrawi AT: Purification and molecular properties of the acetylcholine receptor from Torpedo electroplax. Arch Biochem Biophys 159:362, 1973.
28. Steinbach AB: Alteration by xylocaine (Lidocaine) and its derivatives of the time course of the endplate potential. J Gen Physiol 52:144, 1968.
29. Kordaz M: The effect of procaine on neuromuscular transmission. J Physiol (Lond) 209:698, 1970.
30. Deguchi T, Naharashi T: Effects of procaine on ionic conductances of end-plate membranes. J Pharmacol Exp Ther 176:423, 1971.
31. Maeno T, Edwards C, Hashimura S: Difference in effects of end-plate potentials between procaine and lidocaine as revealed by voltage-clamp experiments. J Neurophysiol 34:32, 1971.
32. Albuquerque EX, Barnard EA, Chin TH, Lapa AJ, Dolly JO, Janssen SE, Daly J, Witkop B: Acetylcholine receptor and ion conductance modulator sites at the murine neuromuscular junctions: Evidence from specific toxin reactions. Proc Natl Acad Sci USA 70:949, 1973.
33. Gage PW, McBurney RN, Schneider GT: Effects of some aliphatic alcohols on the conductance change caused by a quantum of acetylcholine at the toad endplate. J Physiol (Lond) 244:409, 1975.
34. Ariens EJ: Affinity and intrinsic activity in the theory of competitive inhibition. I. Problems and theory. Arch Int Pharmacodyn 99:32, 1954.
35. Ariens EJ, Simonis AM, Van Rossam JM: Drug-receptor interaction: Interaction of one or more drugs with one receptor system. In Ariens EJ (ed): "Molecular Pharmacology," Vol I, Section IIA. New York: Academic Press, 1964, p 11.

194   Maelicke and Prinz

36. Stephenson RP: A modification of receptor theory. Br J Pharmac Chemother 11:379, 1956.
37. Jenkinson DH: The antagonism between tubocurarine and substances which depolarize the motor end-plate. J. Physiol (Lond) 152:309, 1960.
38. Changeux J-P, Podleski TR: On the excitability and cooperativity of the electroplax membrane. Proc Natl Acad Sci USA 59:944, 1968.
39. Hill AV: The mode of action of nicotine and curari determined by the form of the contraction curve and the method of temperature coefficients. J Physiol (Lond) 39:361, 1909.
40. Werman R: An electrophysiological approach to drug-receptor mechanisms. Comp Biochem Physiol 30:997, 1969.
41. Monod J, Wyman J, Changeux, J-P: On the nature of allosteric transitions: A plausible model. J Mol Biol 12:88, 1965.
42. Karlin A: On the application of "a plausible model" of allosteric proteins to the receptor for acetylcholine. J Theor Biol 16:306, 1967.
43. Changeux J-P, Thiery J, Tang Y, Kittel C: On the cooperativity of biological membranes. Proc Natl Acad Sci USA 57:335, 1967.
44. Karlin A, Bartels E: Effects of blocking sulfhydryl groups and of reducing disulfide bonds on the acetylcholine-activated permeability system of the electroplax. Biochim Biophys Acta 126:525, 1966.
45. Karlin A, Winnik M: Reduction and specific alkylation of the receptor for acetylcholine. Proc Natl Acad Sci USA 60:668, 1968.
46. Ben-Haim D, Dreyer F, Peper K: Acetylcholine receptor: Modification of synaptic gating mechanism after treatment with a disulfide bond, reducing agent. Pflugers Arch 355:19, 1975.
47. Kasai M, Changeux J-P: In vitro excitation of purified membrane fragments by cholinergic agonists. I. Pharmacological properties of the excitable membrane fragments. J Membr Biol 6:1, 1971.
48. Kasai M, Changeux J-P: In vitro excitation of purified membrane fragments by cholinergic agonists. II. The permeability change caused by cholinergic agonists. J Membr Biol 6:24, 1971.
49. Cash DJ, Hess GP: Quenched flow technique with plasma membrane vesicles: Acetylcholine receptor-mediated transmembrane ion-flux. Anal Biochem 112:39, 1981.
50. Cash DJ, Aoshima H, Hess GP: Acetylcholine-induced cation translocation across cell membranes and inactivation of the acetylcholine receptor: Chemical kinetic measurements in the millisecond time region. Proc Natl Acad Sci USA 78:3318, 1981.
51. Neubig RR, Cohen JB: Permeability control by cholinergic receptors in Torpedo post-synaptic membranes: Agonist dose-response relations measured at second and millisecond times. Biochemistry 19:2770, 1980.
52. Moore HP, Hartig PR, Wu WC, Raftery MA: Rapid cation flux from Torpedo californica membrane vesicles: Comparison of acetylcholine receptor enriched and selectively extracted preparations. Biochem Biophys Res Commun 88:735, 1979.
53. Aoshina H, Cash DJ, Hess GP: Mechanism of inactivation (desensitization) of acetylcholine receptor. Investigations by fast reaction techniques with membrane vesicles. Biochemistry 20:3467, 1981.
54. Walker JW, McNamee MG, Pasquale E, Cash DJ, Hess GP: Acetylcholine receptor inactivation in Torpedo californica electroplax membrane vesicles—Detection of 2 processes in the millisecond and second time regions. Biochem Biophys Res Commun 100:86, 1981.
55. Anholt R, Lindstrom J, Montal M: Functional equivalence of monomeric and dimeric forms of purified acetylcholine receptors from Torpedo californica in reconstituted lipid vesicles. Eur J Biochem 109:481, 1980.

56. Anholt A: Reconstitution of acetylcholine receptors in model membranes. Trends Biochem Sci 6:288, 1981.

57. Doster W, Hess B, Watters D, Maelicke A: Transitional diffusion coefficient and molecular weight of the acetylcholine receptor from Torpedo marmorata. FEBS Lett 113:312, 1980.

58. Devreotes PN, Fambrough DM: Acetylcholine turnover in membranes of developing muscle cells. J Cell Biol 65:335, 1975.

59. Fulpius BW, Miskin R, Reich E: Antibodies from myasthenic patients that compete with cholinergic agents for binding to nicotinic receptors. Proc Natl Acad Sci USA 77:4326, 1980.

60. Maelicke A, Fulpius BW, Reich E: "Biochemical Aspects of Neurotransmitter Receptors." Handbook of Physiology section I, part I. Bethesda Maryland: American Physiological Society, 1977, p 493.

61. Maelicke A, Prinz H: Preservation of functional properties of the acetylcholine receptor after solubilization. In Schweiger HG: "Cell Biology." Heidelberg: Springer-Verlag, 1980, p 707.

62. Miskin R, Easton TG, Maelicke A, Reich E: Metabolism of acetylcholine receptor in chick embryo muscle cells: Effects of RSV and PMA. Cell 15:1287, 1978.

63. Fambrough DM: Control of acetylcholine receptors in skeletal muscle. Physiol Rev 59:165, 1979.

64. Fu JL, Donner DB, Moore DE, Hess GP: Allosteric interaction between membrane-bound acetylcholine receptor and chemical mediators: Equilibrium measurements. Biochemistry 16:678, 1977.

65. Selikson I, Gibson RE, Eikelman WC, Reba RC: Calculation of binding isotherms when ligand and receptor are in different volumes of distribution. Anal Biochem 108:64, 1980.

66. Eldefrawi ME, Eldefrawi AT, Mansour NA, Daly JW, Withope B, Albuquerque EX: Acetylcholine receptor and ionic channel of torpedo electroplax: Binding of perhydrohistrionicotoxin to membrane and solubilized preparations. Biochemistry 17:5474, 1978.

67. O'Brian RD, Gibson RE: Conversion of high affinity acetylcholine receptor from Torpedo californica electroplax to an altered form. Arch Biochem Biophys 169:458, 1975.

68. Fels G, Wolff E, Maelicke A: Equilibrium binding of acetylcholine to the membrane-bound acetylcholine receptor. Eur J Biochem, 127:31, 1982.

69. Neubig RR, Cohen JB: Equilibrium binding of [^{3}H]-tubocurarine and [^{3}H]acetylcholine by Torpedo postsynaptic membranes: Stoichiometry and ligand interactions. Biochemistry 18:5464, 1979.

70. Weber M, Changeux J-P: Binding of ^{3}H α-toxin to membrane fragments from Electrophorus and Torpedo electric organs. II. Effect of cholinergic agonists and antagonists on the binding of the tritiated α-neurotoxin. Mol Pharmacol 10:15, 1974.

71. Cohen JB, Weber M, Changeux J-P: Effects of local anaesthetics and calcium on the interaction of cholinergic ligands with the nicotine receptor protein from Torpedo marmorata. Mol Pharmacol 10:904, 1974.

72. Changeux J-P, Sugiyama H: Interconversion between different states of affinity for acetylcholine of the cholinergic receptor protein from Torpedo marmorata. Eur J Biochem 55:505, 1975.

73. Schiebler W, Lauffer L, Hucho F: Acetylcholine receptor enriched membranes: Acetylcholine binding and excitability after reduction in vivo. FEBS Lett 81:39, 1977.

74. Raftery MA, Bode J, Vandlen R, Michaelson D, Deutsch J, Moody T, Ross MJ, Strand RM: Structural and functional studies of an acetylcholine receptor. In Sund H, Blauer G (eds): "Protein-Ligand Interactions." Berlin: Walter de Gruyter, 1975, p 328.

75. Boyd ND, Cohen JB: Kinetics of binding of [^{3}H]acetylcholine and [^{3}H]carbamylcholine to Torpedo postsynaptic membranes: Slow conformational transitions of the acetylcholine receptor. Biochemistry 19:5344, 1980.

76. Boyd NB, Cohen JB: Kinetics of binding of [³H]acetylcholine to Torpedo postsynaptic membranes: Association and dissociation rate constants by rapid mixing and ultrafiltration. Biochemistry 19:5353, 1980.
77. Satur V, Raftery MA, Martinez-Carrion M: Propidium as a probe of acetylcholine receptor binding sites. Arch Biochem Biophys 184:95, 1977.
78. Cohen JB, Changeux J-P: Interaction of a fluorescent ligand with membrane-bound cholinergic receptor from Torpedo marmorata. Biochemistry 12:4855, 1973.
79. Martinez-Carrion M, Raftery MA: Use of a fluorescent probe for the study of ligand binding by the isolated cholinergic receptor of Torpedo californica. Biochem Biophys Res Comm 55:1156, 1973.
80. Grünhagen H-H, Changeux J-P: Studies on the electrogenic action of acetylcholine with Torpedo marmorata electric organ. IV. Quinacrine: A fluorescent probe for the conformational transitions of the cholinergic receptor protein in its membrane-bound state. J Mol Biol 106:497, 1976.
81. Grünhagen H-H, Changeux J-P: Studies on the electrogenic action of acetylcholine with Torpedo marmorata electric organ. V. Qualitative correlation between pharmacological effects and equilibrium processes of the cholinergic receptor protein as revealed by the structural probe quinacrine. J Mol Biol 106:517, 1976.
82. Eldefrawi ME, Eldefrawi AT, Wilson DB: Tryptophan and cysteine residues of the acetylcholine receptors of Torpedo species. Relationship to binding of cholinergic ligands. Biochemistry 14:4304, 1975.
83. Bonner R, Barrantes JF, Jovin TM: Kinetics of agonist-induced intrinsic fluorescence changes in membrane-bound acetylcholine receptor. Nature 263:429, 1976.
84. Maelicke A: Kinetics of ligand binding to the nicotinic acetylcholine receptor and acetylcholinesterase. In Burgen ASV, Roberts GCK (eds): "Topics in Molecular Pharmacology." Amsterdam: Elsevier, 1981, p 1.
85. Maelicke A, Fulpius BW, Klett RP, Reich E: The acetylcholine receptor: Responses to drug binding. J Biol Chem 252:4811, 1977.
86. Changeux J-P: The acetylcholine receptor: An "allosteric" membrane protein. Harvey Lectures 75:85, 1981.
87. Prinz H, Maelicke A: Interaction of cholinergic ligands with the purified acetylcholine receptor protein. I. Equilibrium binding studies. J Biol Chem (in press).
88. Prinz H, Maelicke A: Interaction of cholinergic ligands with the purified acetylcholine receptor protein. II. Kinetic studies. J Biol Chem (in press).
89. Jürss R, Prinz H, Maelicke A: NBD-5-acylcholine: Fluorescent analog of acetylcholine and agonist at the neuromuscular junction. Proc Natl Acad Sci USA 76:1064, 1979.
90. Heidmann T, Changeux J-P: Fast kinetic studies on the interaction of a fluorescent agonist with the membrane-bound acetylcholine receptor from Torpedo marmorata. Eur J Biochem 94:255, 1979.
91. Heidmann T, Changeux J-P: Fast kinetic studies on the allosteric interactions between acetylcholine receptor and local anesthetic binding sites. Eur J Biochem 94:281, 1974.
92. Neumann E, Chang HW: Dynamic properties of isolated acetylcholine receptor protein: Kinetics of the binding of acetylcholine and Ca ions. Proc Natl Acad Sci USA 73:3994, 1976.
93. Quast V, Schimerlik M, Raftery MA: Stopped-flow kinetics of carbamylcholine binding to membrane bound acetylcholine receptor. Biochem Biophys Res Commun 81:955, 1977.
94. Dunn SM, Blanchard SG, Raftery MA: Kinetics of carbamylcholine binding to membrane-bound acetylcholine receptor monitored by fluorescence changes of a covalently bound probe. Biochemistry 19:5645, 1980.

95. Barrantes FJ: Agonist-mediated changes of the acetylcholine receptor in its membrane environment. J Mol Biol 124:1, 1978.
96. Eigen M: Diffusion control in biochemical reactions. In Kursunoyla B, Mintz SL, Widmayer SM (eds): "Quantum Statistical Mechanisms in the Natural Sciences." New York: Plenum Press, 1974, p 37.
97. Prinz H: On the interpretation of equilibrium binding studies. J Receptor Res (submitted).
98. Colquhoun D: In O'Brien RD (ed): The Receptors, A Comprehensive Treatise." New York: Plenum Press, Vol 1, 1979, p 93.
99. Sine SM, Taylor P: The relationship between agonist occupation and permeability response of the cholinergic receptor revealed by bound cobra α-toxin. J Biol Chem 255:10144, 1980.
100. Lester HA, Krouse ME, Nerbonne JM, Wassermann NH, Erlanger BF: Light activated compounds as probes for nicotinic acetylcholine receptors. In Birdsall N (ed): "Drug Receptors and Their Effectors." London: MacMillan, 1981.
101. Prinz H, Jürss R, Maelicke A: On the kinetics of cholinergic excitation. Neurochem Int 2:251, 1980.

Index